SpringerBriefs in Applied Sciences and Technology

PoliMI SpringerBriefs

More information about this series at http://www.springer.com/series/11159
http://www.polimi.it

Raffaele Pe

Agogic Maps

From Musical Phrasing to Enhancement of Urban Spaces

POLITECNICO
DI MILANO

Springer

Raffaele Pe
Politecnico di Milano
Milan
Italy

ISSN 2191-530X   ISSN 2191-5318   (electronic)
SpringerBriefs in Applied Sciences and Technology
ISSN 2282-2577   ISSN 2282-2585   (electronic)
PoliMI SpringerBriefs
ISBN 978-3-319-48304-7   ISBN 978-3-319-48306-1   (eBook)
DOI 10.1007/978-3-319-48306-1

Library of Congress Control Number: 2016962041

Printed on acid-free paper

This Springer imprint is published by Springer Nature
The registered company is Springer International Publishing AG
The registered company address is: Gewerbestrasse 11, 6330 Cham, Switzerland

Foreword: Tune-In City

The city is the problem; that much, we know. Since Tafuri's demystification of the logic of metropolitan architecture in the late 1960s, in which the collusion between economic and cultural forces was denounced, architects have increasingly found difficulties in asserting their role to find social and political traction for their ideas. To be sure, Tafuri was not responsible for opening an unbridgeable chasm between city and architecture; rather he should be credited for lucidly detecting and mapping it. The shockwaves of this realization are still legible today, as it suffices to have a quick look at the shelves of architecture sections in bookstores to register the flurry of trends, full-fledged theories, or quick provocations still agitating the debate in architecture and urbanism. Likewise, disciplinary insecurity has led architects to wander into both adjacent and distant fields: landscape, computation, science fiction, political theory—to name a few—have been utilized to provide new blood to heal the Tafurian fraction.

The research carried out on agogic maps by Raffaele Pe, which is well reconstructed in this book, returns to this difficult and yet crucial problem providing a new opening to understand and successively intervene in the city. He does so by establishing two key themes we will dwell on in this short text. First of all, there is the choice of the city in which to intervene: Dar es Salaam—Tanzania' capital and a testing site for an 'agogic' mapping exercises—poses problems that are far from those traditionally encountered in Western cities, but by no means tangential. Whereas much of avant-garde design theory has been focusing on injecting dynamism, if not altogether chaos, into carefully planned Western cities, Dar es Salaam, on the contrary, is in need of order, of a syntax to encode its complex and unregulated growth. Rapid urbanization, deprived conditions, and yet a rich local culture and tradition have not organically blended, not settled, nor have they given rise to a pacifying image. The question of order thus becomes a complex one in which traditional, imported, if not colonizing, models cannot be proposed.

This latter consideration appears to be the starting point for the second element of novelty provided by this piece of research. Agogic mapping draws from a variety of incredibly interesting and yet new sources of inspirations and reflections for architects. Not only music—from which it directly comes—but also scenography and semantics are all employed to conjure up a series of new and known techniques to map urban environments. It is important to point out Pe's background here as he splits his time between his musical and architectural careers; his knowledge of both disciplines is direct, based on actual practice. As suggested earlier, this biographical note distances him from other architects whose forays into other disciplines have been characterized by superficial knowledge and, consequently, less-than-fortunate results. This is not the case here as the range of specific and original musical references clearly confirms. It is a pleasure to be taken through the works of Appia or Donatoni in what we found the most interesting part of the book.

Agogic thinking is virtually absent from architectural theory, as well as from the repertoire of representational techniques available to architects. As such it deserves some further contextualization in order to fully appreciate its original contribution and potential to alter our reading of cities. The second half of the twentieth century saw a number of attempts in architecture to conceive of uncertainty not as a lack, a weakness, but rather as a fertile condition for design innovation. The gap between event and space was perceived by Bernard Tschumi and Nigel Coates in the early 1970s in their work at the Architectural Association (Tschumi 1983) as a source of inspiration for narrative architecture. Not long after Stan Allen began his inspection of art practices to produce not only his *Field Condition* (1985), but also, more relevant in this context, his *Barcelona Manual* (1996), a completion entry for the Logistical Activities Zone of the Catalan city was delivered in the form of a succinct manual. Here, too, we find a productive gap between script and performance, a gap for the designer and users alike to inhabit. Agogic mapping techniques promise to expand this space, opening it up to more ephemeral, perceptual dimensions, enriching it further to reveal a virtual, but by no means, less tangible image of the city.

Finally, we would like to conclude with a brief reference to digital technologies and their role in mapping. Raffaele Pe elegantly integrates elements of local morphology and "agogic theory" in his computational models to generate multiple maps of Dar es Salaam. The role of digital technologies in this endeavor is both essential and limiting. Obviously, the complexity of factors considered, the dynamic nature of agogic maps, and the morphology of informal settlements all lend themselves well to the possibilities to manage large data sets enabled by CAD software. In this sense, the research mostly shows the potential of this way of operating, calling for more advance experimentation in this area. However, by combining quantifiable elements with more elusive ones, this research challenges the limits of what can be computed and how it relates to urban dynamics. This part of the research too hits on a 'sweet spot' of computational discourse in architecture and would deserve more scholarly attention and experimentation. What is at stake, however, is already clear in Raffaele Pe's research: the possibility to unearth a sort of unconscious dimension of the city, a virtual—defined here in Deleuzian terms as the opposite of actual—through which to attempt, once again, to relate the city to its architecture.

How will the city be described by the conflation of digital affordance and musical understanding? What pedagogical lessons can be transferred from Dar es Salaam? And where to apply them? These are some of the questions elicited by reading Pe's work. These issues are central to contemporary architecture and urbanism; they call on students, designers, and theoreticians to take them seriously and engage them as design opportunities to map the urban, social, cultural, political, and ecological implications.

Roberto Bottazzi
London, UK

Foreword: Unconventional Media for Metropolitan Orientation

2019 Los Angeles. The consequences of dystopias are visible all over the planets. An oppressive and hostile environment justifies the migration of many humans to the otherworldly colonies using cars, which fly all over the sky following random paths.

This is the future as described by Ridley Scott represented in the movie *Blade Runner* in 1982.

Today (2016), our cities which have become metropolises that are not able anymore to reconstruct a link between the body of their inhabitants and the body of space, as Raffaele Pe argues in this book, neither the innermost expressivity that manifests a performative exchange between houses and routes.

The problem of how the old parts of the city or the new and neglected ones are integrated in the whole is emerging.

Many other questions arise. For example, how is it possible to design buildings or urban areas to sensitively respond both to the physical content and, at the same time, so that they can distinguish themselves from it in the form of a global/local landmark?

It is necessary to think about a new language of composition that refuses direct references to a pseudo-picturesque historicism. Thus, the public realm requires the construction of a narrative through our architectures that tells stories about who we are and what a city wants to be. It introduces above all a symbolic dimension into the architectural project using formal archetypes that are able to evoke a new meaning in the global culture.

Now, an intentional and naïve removal of the anthropological time of the physical space, such as the fundamental basis of common sense and citizen participation, is a matter of fact. We should question ourselves wondering if it is necessary to enjoin scientific, technical, and political rationality in order to take part in modern civilization. This is something that very often requires the pure and simple abandon of a whole cultural past. It is a fact that every culture cannot sustain and absorb the shock of modern civilization. So here is the paradox: how do we become modern while returning to sources, how could we revive an old, dormant civilization while being part of a universal civilization? (Ricoeur 1961).

A mapping project is able to represent the new metropolitan complexity of a city with the aim to reconnect the dimension of the metropolis with a real and memorable experience. Maps can reproduce a representation of the metropolitan hybrid space composed through relevant points, since they foster a deep knowledge of the metropolitan structure and the relationships that tie together its nodes in time intervals. This connection is possible only through an intense experience of the time-space. Therefore a map should show how every metropolitan phenomenon is constituted by a series of feelings related to the intense experience of the place. Art is the key to create memorable places.

In this respect, Raffaele Pe suggests that a map does not only describe the organization of objects in space, but it is also a tool to reveal and experience a built environment. Outlining a map is not only a matter of drawing, it is also a matter of physical perception, and it represents an issue of orientation, a permanent bond between the body of the territory and the body of its inhabitants. We acknowledge mapping as a practice of spatialization: not only does it implicate the geometric measurement of things, but it also refers to the reported, tangible effect that such measures have on the body of those who experience them.

A Technological Issue

New devices define a city as a *meta-space* (Bunschoten 2003). A city is a fluid form of public spaces that evolves over time, generating different definitions of public realms and different ways of participating in it. Meta-spaces enable bringing dynamic scenarios into the flows of a "second skin". As presented by urban theorist Grahame Shane, a meta-space in the second skin is a public space, a public matrix, that we name Meta-city (Shane 2005).

A research on new metropolitan territories must critically reflect on the exploitation of digital technologies and media as specific tools for the collection, organization, and interpretation of data for city analysis and architectural composition. Aesthetics is here proposed as a language across different disciplines, a tool that is able to support creative and innovative ways of designing and ruling future stages of development of the city. Architecture among the other arts can be a way to configure a different vision on contemporary metropolises and our diverse society: it is a design instrument for reactivating some parts of the city. With the use of new technologies applied to the context, Pe wants to determine some sensory inputs causing specific memories of a particular situation, hitting the senses of the observer: this book conceptualizes and spatializes feelings to turn them into a conscious perception, into a knowledge form that transforms a traditional spatial experience through the movement of the body.

With the idea of agogic maps, the author of this book wants to produce an evocative image, the perfect place from a symbolic point of view on the city. This is, for the MSLab research unit of the Department of Architecture and Urban Studies at Politecnico di Milano, a way to reactivate urban cultures producing new synergies. The symbolic is deeply rooted within a particular culture domain, and it indicates the presence of a topographic identity. This, in particular, is our Italian and Milanese approach to the dialogue among cities.

An Issue of Interpretation

With this book, Raffaele Pe offers a brilliant contribution to the research on the generative process of urban forms. This text links urban studies to a different way of understanding the harmony of the city. Pe finds musicality even in the dissonances of the city: he understands the dissonance as a form of complex harmony after the enlightening introduction of Arnold Schönberg's introduction to the topic. The final goal is the conception of a new notational system called agogic, in order to establish a meaningful relationship between settlement pattern and the continuous variation of the anthropological image. These new tools of analysis, interpretation, and design become indispensable in unstable contexts or within those that are provided with a weak identity.

This contribution to this field of research is not only related to musical-performance parameters from a technical point of view. The study considers also the orientation process in sensitive areas, which is not only a technical and material method but also an intellectual one. Raffaele Pe's mapping process integrates constituent processes of spatial configuration and architectural structures (ecologies); his application of the method to the context of Dar es Salaam is quite clear. This fact is especially important for compositional studies of urban development in areas relevant from an environmental point of view: cultural landscapes in developing countries.

Pe focused his study mainly on the Theory of Form discussed through a study which runs through post-structuralist criticism, generative semiotics, and ethno-semiotics to the field of spectralism and psychoacoustics, which is new to our approach to urban studies. Spectralism and psychoacoustics are conceived as fundamental tools for the definition of a spatial narrative made of sounds. The sound event is therefore considered an essential medium of knowledge to shape the space due to the fact that it is an immediate meaning producer. Against a purely chronological/causal and type-morphological reading of the urban environment, this approach has favored an experimental study aimed at understanding the constitutive processes of a contemplative environment. This approach has an important effect on the analysis and the project of the contemporary city because new metropolitan territories are normally analyzed with the intention of revision. Agogic maps also deal with issues of the relationship between architecture and society, such as local identity and social and cultural sustainability of globalization, however it is a symbolic and imaginative approach that transforms the configuration of the meta-city through performative actions.

The proactive attitude of the author denotes intellectual vitality in the various topics discussed. He conveys an original insight that is central to the construction of the method of analysis, especially in non-European contexts. The research is part of the architectural culture study of the MSLab—Politecnico di Milano, and it analyzes the current urban design theory across both European and international literatures with a special emphasis on contemporary musical composition. We should not forget that Raffaele Pe is a musician himself, very active in the rediscovery of the ancient repertoire, employing past techniques with a modern sensibility.

Finally, this study deepens design procedures that take into account new digital design technologies. This enables creating original critical experiments on the basis of interdisciplinary analogies revised in an through a satisfactory theory. The volume extrapolates concepts already discussed in Anglo-Saxon countries, but perhaps not yet analyzed inside the context of Italian schools of architecture. The book critically introduces new terms for discussion such as the contrast between catastrophic and morphological continuity of deformation (trans-morphosis), the redefinition of the idea of the structure that exceeds the archetype of object and subject, the elaboration of an abstract pattern that feels more and more the need for an alternate configuration evaluating new geometrical systems of description.

The final research question is how a map can introduce a genetic perspective and a generative space based on the topographic variations of the settlement? Could the exploitation of sounds in the way we "write" the space affect the curvature radius of a settlement model? The climax of the research leads to a syntax of the spatial form for the development of a design methodology, integrating objects within their environment through digital technologies, an original practical and theoretical contribution even though still in embryo. This is an original work which perhaps will receive broader specifications in future projects for the definition of contemporary processes of logical succession in sense-making for metropolitan orientation.

Antonella Contin
Milan, IT

Acknowledgements

Completing this book represents both the end of 5 years of research, as well as the beginning of new possible horizons in this field of study. This would have not been possible without the esteemed lead of Ernesto D'Alfonso, Professor Emeritus in Architectural Composition at Politecnico di Milano, and Antonella Contin, director of MSLab, School of Architecture, Department of Architecture and Urban Studies (DASTU), Politecnico di Milano.

I would like to thank also Massimo Della Rosa for his precious contribution concerning historical and morphological studies in Dar es Salaam, partly included in *New Models in Planning Practice to Address Migrations in the Sub-Saharan Region* (Della Rosa 2013). The prototype of Dar es Salaam Agogic Map is available online (Geoscore), and it was created with the support of Alessandro Musetta and Stefano Bovio with whom, in 2015, we launched Sound of Things, a permanent research lab on architecture and sound studies (www.soundofthings.org).

Finally, I would like to mention music theorist Diego Fratelli, professor at Civica Scuola di Musica Claudio Abbado, Milano, and composer Martino Traversa for their irreplaceable guidance in exploring the world of ancient and contemporary music for a meaningful and profitable exchange with architecture and spatial design.

Contents

Chapter 1
The Imageless City. An Aural Approach to Locative Media and Informal Settlements

Abstract How could musical knowledge constitute an asset in configuring urban spaces in the current technological context? This book tries to answer this question deepening the relationships between urban morphology, locative media, and sound design, with a special emphasis on Marshall McLuhan's definition of the Acoustic Space, the performative ambience emanated by modern technologies. This book investigates the case study of the African capital of Dar es Salaam in Tanzania though this approach. This case study provides an example that is far enough from our conventional understanding of urban structures, a conurbation with uneven characteristics that challenges and criticizes the extent of common architectural solutions to envision the future development of the notion of metropolitan livability in a global frame.

1.1 The Experimental Extent of a Map

The Dar es Salaam map prototype illustrated in this book interrelates emerging topographic qualities of the built environment with site-specific sound material to orient and determine feasible spatial movements. In order to validate these correspondences, the map exploits traditional surveying outlines, adding a new layer of meaning onto these geographic and architectural signs. According to this research, the spontaneous aspect of informal territories expresses a value of openness and variability towards more imaginative and performative ways of describing our habitat through maps. The fascinating morphology of Dar es Salaam declares, with more effectiveness, its habitability and spatial practicality because here the built environment intertwines nature and virtues, geographic advantages and technological development, a condition that the modernity of the globalized world seems to discourage.

In accordance with Kevin Lynch's belief for which "city design is a temporal art [...] Moving elements in a city, and in particular the people and their activities, are as important as the stationary physical parts" [1], the case study of Dar es Salaam demonstrates that this urban landscape is a moving object transforming at a much faster speed than what we are normally used to in consolidated settlements. This approach allows for a broader investigation into the notion of modernity and the relevance of the environmental extent of the signs we apply to the description of our habitat.

The prototype illustrated in the book frames the expanse of the proto-urbanized territory of Dar es Salaam, designating the city as a paradigm of a dynamic settlement, which means the formal expression of the rhythm impressed on the architecture of the city by the spatial behavior of its inhabitants and the climatic mutations of the environment. The city shape of Dar es Salaam is the resulting common space in between private houses and metropolitan infrastructures. This space is built informally and spontaneously, however it is widespread over the whole territory, with sporadic and often illegal connection with the formal network of metropolitan infrastructure such as motorways, railways, electric grid, and the sewage system. Dar es Salaam is the image of any city before actually becoming a city, according to a "modern" acceptance: this African conurbation has not yet disjointed its physical configuration from the vernacular and natural instances emerging from the spatial behavior of those who have created and transformed such a place. The agogic map of Dar es Salaam is a sense-making device that emphasizes this condition and stresses the tactile extension of spatial writing through modern technology in order to improve practicability.

The map emerges from the relationship between resilient and spontaneous elements of these urban commons. The device fixes the objects that characterize, in a more permanent way, the urban landscape and manages, at the same time, the elements that transform this geography with peripatetic alternations. The map consolidates what is spontaneous, indexing its nature in any of its variations. The map reconnects the elements of the built environment with the origins of its morphology, as Vidal de la Blanche would say "the boat glides on the water surface, the waves settle and the groove is deleted; the Earth is more faithful as it preserves the footprints of the routes of the early-rising workers. The route is impressed in the soil and it spreads seeds of life such as houses, villages, cities" [2]. Houses and routes are intertwined in a relationship that expresses well the existing link between migratory and settling forms.

A map does not only describe the organization of objects in space, it is also a tool to reveal and experience an environment. Outlining a map is not only a matter of drawing, it is also a matter of physical perception: it represents an issue of orientation, a permanent bond between the body of territory and the body of its inhabitants. The map constructs a "bounce" [3]—as music composer Franco Donatoni would suggest—between the antecedent and the consequence of this condition: the territorial extent of the habitat and its geographic representation. The cognitive bounce of an agogic map produces a passage from reality to space, and, vice versa, from geography to nature. From this perspective, we understand mapping as a practice of spatialization: not only does it implicate the geometric measurement of things, but it refers also to the tangible, reported effect that such measures have on the body of those who experience them.

Such cognitive bounce between the resilient and moving objects of our built environment focuses also on the transition between city and countryside, between urban and rural, requiring—especially for Dar es Salaam—a special definition of this condition in those constructions that are still under development in informal areas. In these areas, this situation arises with greater clarity, as in these neighborhoods natural and rural elements are very present and more often they actually provide the condition for the proliferation of uncompleted rustic buildings. Here, ordinary forms of urban planning are not able to control such proliferation as the settlements follow morphological rules that fall outside the principle of efficiency and rationality typical of a contemporary formal metropolis.

The space of a map, as a cognitive means to measure the distance between our body and the external reality, is endemically constituted by moving and unstable elements which are as relevant as the stationary physical references to orient our movements. Urban spaces especially today are not able to manifest their meaning and their function with clarity due to a weakness of their linguistic and syntactical/architectural equipment. Moving and stationary elements both contribute to shape of the image of the city, which is often a most effective cognitive instrument to memorize its articulation and the quality of its inhabitable enclaves. On one hand, the image of the city embodies the lineaments of its built environment, while, on the other hand, it retains the elusive traces of recurrent spatial behaviors.

1.2 A Topographic Bond: Houses and Routes in Informal Settlements

Migrations naturally tend to settle in the place where natural resources are harvested and shared through various economic and societal exchanges, on the basis of criteria like availability and accessibility. "The archetype we refer to is the one for which the appearance of civilization (from the Latin word *civitas*, which means city) is the result of a sequence that starts with the collection of spontaneous fruits and wild game: it proceeds with agriculture, and it culminates with the formation of the urban environment of the state. This is a linear sequence, continuous and cumulative, that has been argued only from the second half of the 20th century, and that is divided into phases, the incremental steps of development of humankind" [4]. The route, which is a moving and performative element, is always the sign of the presence of a house that stands at its end, and it moves along natural topographic trajectories.

Modern infrastructures, with the superimposition of rectangular grids on the natural flow of the soil surface, liberated the route from the constraints of topography, at times renouncing communication of the effective condition of the places in space and time. The tactile power of the map recalls in the mind of the reader the factual quality of the spaces described. We noted how the passage to modernity in terms of spatial writing implied the loss of the experiential in favor of a principle of kinetic efficiency. The omission of the territorial datum from the isotropic map of modern infrastructures decreases the capacity to detect climatic and surface emergencies. Our cities, now metropolises, are not able anymore to reconstruct a link between the body of its inhabitants and the body of space, its innermost expressivity that manifests the performative exchange between houses and routes. "For the ancient Greeks the metropolis indicates the town-mother, and it implies a relationship with a town-daughter, a colony, like the relationship between Athens and Thurii. For them the metropolis is the city where we register the triumph and simultaneously the death of space, intended not only according to its general and technical acceptance, but also for the indexical relation between street and locations" [5].

In irregular unplanned settlements like the ones that characterize the majority of the territory of Dar es Salaam, traditional surveying tools seem to lose their effectiveness in fixing on the map the qualities of a landscape that transforms seasonally. Due to climatic, economic, and cultural agents, the morphology of the settlement changes together with the modification of the place. The most peripheral parts of these settlements are normally the most exposed to these external influxes, therefore those moving objects are usually imperceptible to general cartographies.

1.3   The Notion of Agogic to Evaluate Performance for Spatial Configuration

This book presents the tool of agogic maps as a means to manage the shifting condition of these territories, while identifying a set of updated mapping techniques to orient people in a place that changes its architectural configuration at the time of each flooding season. The informal characteristics of Dar es Salaam are interpreted in this piece of work as an opportunity to apply devices that enhance the performative profile of the city to this topography, both for local and foreign users. The observations of this research, supported by the delineation of agogic maps, lead us to a different conclusion in regards to the imageless appearance of Dar es Salaam. Such a misconception is partly the result of the restricted capacity of current surveying techniques to discern resilient elements from spontaneous ones, in order to reconstruct the variable dynamics of settling and transformation processes.

Agogic maps coordinate the navigation of our body within informal settlements starting from the musical premises of a rhythmical regulation of urban movements. An agogic map is a device for an intercultural dialogue between local inhabitants and external city users, and it aims to open new perspectives, not only for what concern the traditional cartographic writing modes, but also on the language we employ to communicate the urban values of such peculiar spatial conformations. Agogic accents in music represent a rhythmical practice for which the relationships between the interpreter and the score are coordinated to produce variation and expression within the extent of a regular pulsation. The agogic mutation of a homogeneous pulse into an irregular pattern grants the emergence of governable degrees of expressivity in phrasing musical sentences on the basis of poetic meaning or rhetorical instructions.

The agogic of the settling model of Dar es Salaam on the metropolitan scale is looking for a compelling interchange with informal systems. Although the first seems to be more efficient and perhaps more livable than the second, irregular areas demonstrate their sustainability in relation to topographic convenience in regards to sun exposure and primary food sustenance. This African case study is designated as a paradigm for an ampler series of cases. Dar es Salaam shows not only its relevance to our concerns regarding the consolidation and the stabilization of the urban sprawl of our metropolises, but it also refers to other urbanized areas with similar issues. Mapping informal settlements in Dar es Salaam enables a deeper reflection on problems of spatial disorientation in unsettled built environments and creates new possible fields of study concerning inter-scale compatibility between adjacent, though separate, urban elements. Agogic maps outline the emerging relationship between places, spatial behavior, and settling forms.

1.4   Tonal Transcripts

Which writing mode should be more appropriate to describe the shifting elements of contemporary urban landscapes? Could a tonal system, informed on the basis of musical principles, constitute a reference for the geographic outline of such a settling context? In Dar es Salaam, accessibility to informal residential areas is particularly difficult due to the presence of an element of degradation and decay diffused throughout the urban territory. On the other hand, the evasive, fleeting character of irregular settlements often guarantees their survival and protects the life of their inhabitants. If we look closely at the shape of spontaneous settlements, we could actually discover a number of assets related to these spatial configurations because they manifest a more effective integration with the existing topography.

Issues of cartographic rendition of the informal urbanity of Dar es Salaam are often interconnected with the impossibility to link minor architectural elements belonging to informal districts with larger infrastructural units at the scale of the metropolis. Such scalar incompatibilities require a greater effort to envisage new urban devices to facilitate a proficient exchange between informal agglomerates and the main metropolitan infrastructures in order to ensure, on one hand, the betterment of the living conditions of irregular settlement and, on the other hand, the morphological evolution of the eco-tonal lines of the city in constructive forms that accept the performance of moving elements as a fundamental compositional equipment of modern built environments.

The agogic map of Dar es Salaam constitutes a pre-design tool that considers all these circumstances as part of a process of urban configuration. The map is a "low-definition" medium, a device to help users to imagine and create their own sense of affiliation to the place while navigating safely through informal, shifting urban landscapes. The implication of a locative digital apparatus allows for a dynamic management of the cycles of climatic transformation of the place emphasizing the topographic expressivity of these areas. This prototype intends to provoke the cognitive bounce indicated by Donatoni, starting from the territory towards its virtual transfiguration, and then back, passing from the abstraction of the mental space of the device back to the sensitive experience of the place.

1.5   Investigating Informality in Dar es Salaam Irregular Settlements

Dar es Salaam, due to its maritime and rural origins, and like many cities that do not belong to the Western tradition of urban constitution, can be perceived as an imageless city, an aggregation of built elements without an acknowledged physical appearance in the memory of his inhabitants. This condition is produced by the fact that Dar es Salaam is widely known as probably the most informal of all the African cities: a contemporary metropolis of nearly three million inhabitants, established for more than the 80% of its territory on irregular districts in constant transformation. Because of the informal conditions that determine the formation of such an extraordinary built environment, the urban form of Dar es Salaam has not yet been fixed on a map emerging from a shared action of territorial definition. Traditional cartographic writing does not constitute a meaningful instrument to describe the drifting and fast-changing reality of this city. The mental image of the city, an urban meta-space envisaged by its inhabitants to navigate the built context, needs to exploit alternative and probably unconventional media to achieve such an objective.

The Dar es Salaam prototype insists on environmental agents that shaped that urban landscape through history and the elements that defined its resilience. The emergence of urban spaces in this context corresponds to the enlargement of the original rural settling units, generally centripetal and self-contained within walls that separate them from the fields. This condition coincides with a number of advantageous economic, climatic and cultural aspects, because of which the unit becomes too big to be protected within its normal boundaries. Therefore, from a nuclear shape, the unit develops into a reticular form [6], declaring the passage from a pre-modern agricultural vocation to a modern ambition of urbanity (Fig. 1.1). The map focuses on this specific historical phase, in order to preserve and enhance the meaning of current spatial distributions in their connection with the initial settling behavior. Those spontaneous elements (elements that were configured coherently with the primary needs of sustenance of the population and the territorial features of the place) become a signal to evoke the original meaning of such urban disposition.

Our efforts focus on what was spontaneous in this landscape and suddenly became a means for morphological stabilization in the new distribution of urban elements. Exactly as it happens in the walkable sets of set designer Adolphe Appia,

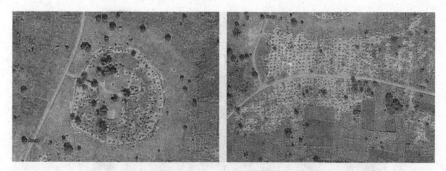

Fig. 1.1 A comparative study of a Ugandan village where both models of urban growth appear. The passage from a circular to a reticular model happens when the circular model is not able to grow beyond a certain width (Google Maps)

where architectural constructions are continuously reorganized on stage during the performance according to the development of the plot, while they preserve the sense of their original shape, built objects in the urban configuration of the city recall the original impression of their form in relation with the reasons of their nuclear institution.

Consistent with the study of the morphology of informal dwelling units in Dar es Salaam, we understand the extent and the trajectories of people's most relevant urban movements. According to geographer Franco Farinelli, the origin of geographic measurements, therefore the emergence of what we call "space", is to be attributed to the invention of the grid: "Equivalent in Greek is expressed with the word "parallel", and the invention of the space has to be considered in relation to the introduction in the description of the Earth of the so-called "geographic grid", or better of the network of parallels and meridians with which we try to reproduce on paper the globe's curvature [...] Cartographic projections are based on mathematic rules that allow to determine a correspondence between a point on the plane surface of the sheet with a specific point on the globe at the intersection of a meridian with a parallel". However, in the urban field of study, we register both fixed points that resist spatial and temporal mutations, together with elements that present a higher degree of mobility in space and time. Geographically moving objects are those elements that spontaneously, in an exceptional and somehow chaotic way, regulate the modulation of the urban system.

1.6   The Field of Action

Dar es Salaam (Fig. 1.2) is an urban conglomeration with a population of more than 3 million inhabitants, of which nearly the 80% live in irregular dwellings raised in spontaneous ways without any planning regulation by the local authority. The population of Tanzania reaches nowadays the number of 35 million people, of

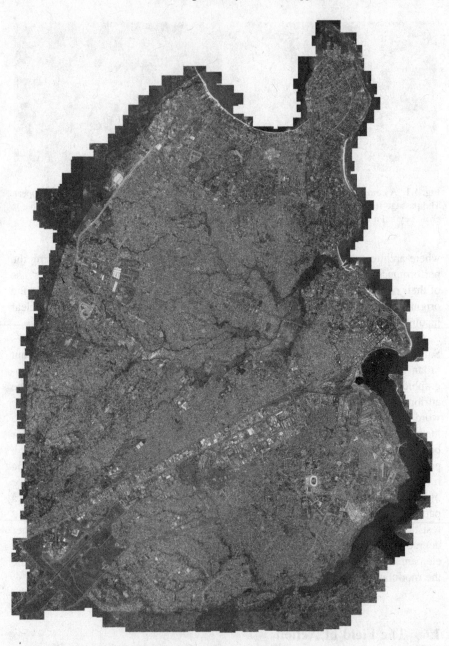

Fig. 1.2 Aerial view of the Urban Region of Dar es Salaam (Google Maps)

which one third lives in cities with a rapidly-developing process of urbanization generated by a phenomenon of mass migration from the rural areas of the country [7]. In order to frame the geographic context of Dar es Salaam, the orographic conformation of Tanzania constitutes a relevant starting point. This land presents important mountain ranges in the northern part of its territory, culminating with Mount Kilimanjaro, the highest peak in Africa (5895 m), while the coastline and the south are generally flat. The southern part of the *Great Rift Valley* overlooks the northern border of Tanzania, containing numerous lacustrine basins of continental relevance, including Lake Victoria and Lake Tanganyika on the western border.

In Tanzania, there are only two urban centers acknowledged as cities: Dar es Salaam and Mwanza. The topography of this region is constituted by various elements: the north-east and south-west side of the urban region is characterized by the presence of hills and small valleys that delimitate the alluvial plane of the coast where the cities spread. The orientation of these two geographical features, of which the most important are the *Pugu Hills* (maximum height of 330 m), expands from Dar es Salaam to Kirasawe towards the south-west. The city region of Dar es Salaam is also enclosed within a natural hydrographic basin originating from the presence of the River Ruvu and River Wami. The hydrologic system within the urban region includes four major perennial rivers, called *Mzinga*, *Kizinga*, *Msimbazi*, and *Mbezi*, and a large number of minor seasonal streams. In this region, the rainy season register two annual peaks, respectively the *Long Rain* in April and the *Short Rain* in October. The city of Dar es Salaam is located at the southern end of the *Great Rift Valley*, on the east coast of Tanzania facing the Indian Ocean between Mafia Island and Zanzibar.

Dar es Salaam takes its name from the Arab tradition, comprising the ethnic majority of the country, and it means "place of peace", literally in Arab "the residence or the harbor of peace". The urban development of Dar es Salaam spanned the first half of the 19th century up until the beginning of the 20th century, gathering together several rural villages located along the coast. The city was founded in 1862 by Sultan Seyyid Majid of Zanzibar, and it was a commercial hub for the exchange of slaves and ivory with India, Europe, and America. Urban growth developed very rapidly, and it became the capital of German East Africa in 1891 by a decree of the German colonists. Between 1917 and 1961, it was chosen as the capital of Tanganyika under the British occupation following World War I, and finally in 1961, when Tanganyika declared its independence from the United Kingdom, Dar es Salaam became the capital of the United Republic of Tanzania, formally constituted in 1964, after unification with Zanzibar.

The growing population of Dar es Salaam since 1964 has been connected with the commercial and infrastructural development of the city. The need for manpower to complete these first important urban transformations fostered the migration of young men from the countryside to the city. Finally, the second wave of population increase was registered during the last thirty years due to the same migratory process from the countryside towards the city, although this time it achieved a more stabilizing dynamic in which entire families and clans moved to the city in an unrestrained way. This process of urbanization determined the expansion of irregular districts, overpopulating existing settlements, and generating new spontaneous irregular dwellings. This condition provoked also a remarkable deterioration of public services and the inevitable evolution of illegal economic activities.

Due to the difficulties that emerged to guarantee occupation to all the immigrants, the degree of urban poverty increased immensely across the whole urban region. Up until the end of the 1980s, in accordance with the agricultural socialism promoted by the government (*ujamaa*), the institutions tried to encourage the poorest part of the population to return to the countryside. In recent years, a notable political and social shift suspended this action, resulting in a significant decrease of the living quality in many urban districts.

Dar es Salaam is built upon a radial model of expansion due to the emergence of suburban areas around 1980, and the construction of unplanned dwellings along the coast and along the major infrastructural arteries toward the north, west and south, namely *Bagamoyo Road*, *Morogoro Road*, *Pugu Road*, and *Kilwa Road* (Fig. 1.3). The urban development was designed according to different plans of transformation, in 1948, 1968, and 1979. All three plans were unsuitable to the management of the population growth, leading to the settling of people coming from rural areas in unstable informal districts. In this phase, the first examples of irregular residential

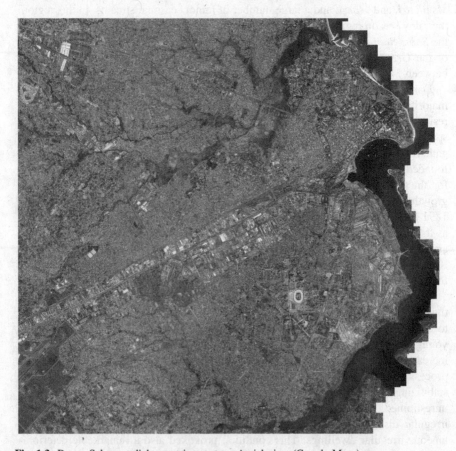

Fig. 1.3 Dar es Salaam radial expansion system. Aerial view (Google Maps)

areas arose in the city—the so-called *squatter areas*—where more than the 70% of the population now resides [8]. There are more than 40 irregular residential districts in Dar es Salaam, most of them with a very high population, with buildings irregularly situated and inadequate hygienic and sanitary services. Also, many informal settlements are located in geographical areas affected by a high degree of hydrological risk, along alluvial valleys and seasonal streams. Dar es Salaam is an open urban system, monocentric and non-rectangular, with a center of gravity located along the coastline. The existing infrastructural network creates a feedback mechanism between the inner formal district and the peripheral areas. In the meantime, small–scale exchanges between formal and informal districts are inhibited. This condition is particularly aggravated by the presence of informal areas contiguous to main commercial and administrative centers, discouraging a transversal crossing within the city. Informal districts, instead of growing outside their boundaries, engender processes of compaction, saturating all the empty spaces in between existing constructions.

Inter-scalar compatibility between different districts emerges as one of the most relevant issues of this metropolitan region. Analyzing its urban morphology, we recognize diverse infrastructural ambitions for the formal and the informal city in relation to various topographic conditions. While the formal, built environment manifests a clear intention to include within the boundaries of the urban region a set of villages belonging to the countryside, informal districts appear as self-contained urban enclaves that are impenetrable and closed to the surroundings: the access to these places can only happen on foot or with light vehicles. Plans provided by local authorities present remarkable omissions in the provision of population growth, especially in these neighborhoods. Moreover, they do not consider the possibility of enabling a thick network of transversal connections between various municipalities and the informal settlements in order to foster accessibility in every part of the urban region. The formal city also lacks in implementing public transportation from consolidated areas towards more peripheral and less resilient suburbs, exaggerating distance and disjunction between the urban parts, and modeling each part in the shape of a compound.

Dar es Salaam' urban fabric within informal districts is rooted in the topographic structure of the soil. The architectural transformation of the place affects the territory in order to improve the growth of site-specific economic activities of sustenance, such as agriculture and manufacturing with local raw materials. Formal and informal settlements present a very different relationship with the surrounding landscape, intended not only as a cultural environment but also as a work space: experiences of urban agriculture are detected within the boundaries of those areas that are more detached from the city center and the coast though contiguous with the countryside. In such a context, the most common work activities are the planting of crops like wheat and coconut, the breeding of cows and goats for the family use, fishing, and sand extraction for building purposes. Great is the care that the inhabitants of the informal settlements afford the natural environment as is shown by their implementation of artificial devices for the protection of mangroves along the coastline, supportive of fishing and the reproduction of the fishes.

The configuration of formal districts in Dar es Salaam shows inefficiency in the management of environmental issues, such as seasonal flooding, with adequate infrastructures. The inundation cycle of the alluvial plain on which Dar es Salaam is built appears twice a year: April (long rain) and October (short rain). Under these circumstances, orographic depressions fill up with water coming from the surrounding highlands and from overflowing rivers and aquifers, which cause the partial inaccessibility of the neighboring settlements [9]. Informal settlements are frequently located along these potentially dangerous ecotonal lines. During the flooding cycles, the destruction of temporary dwellings and shacks is inevitable.

However, at a closer viewing, the morphological disposition of irregular settlements tries to avoid flooding by locating their pedestrian routes on contour curves on higher levels, defining the safe and more resilient dry areas. In this perspective, the configuration of these special urban regions presents only a seeming informality: irregular settling models adhere with greater efficacy to the topographic deviations of the soil, as well as guaranteeing the survival of entire communities during the flooding seasons, and revealing the lack of capability of planned infrastructures at the urban and metropolitan scale to manage such emergencies.

The Ruvu River, in the northern part of Dar es Salaam, constitutes the main water source of the city with a capacity to supply more than three million inhabitants. However, issues of distribution, storage, and water treatment [10] diminish the potential to reach the population on a consistent basis. Numerous are the cases of water pollution due to the presence of industrial plants located along streams and main infrastructural routes. Large-scale food production and food treatment are the main activity of these plants, as well as the production of fertilizers, plus tiles, tanneries, and refined oil. The excessive functional division of the territory in residential, commercial, and productive areas makes it more difficult to assure a capillary distribution of services to every part of the city. Such simplification of settling units outlines stable, though convoluted, built environments that are unable to interchange with each other. The formal city does not seem to be effective in building enduring and coherent integration among all parts and so improving the accessibility of certain areas in relation to emerging activities. The small-scale, spontaneous urban fabric is more hybridized and varied, enhancing its transitional, fast-changing vocation. Finally, the elusive character of informal settlements do not interact with more stable assets provided by the planned city: the lack of social, administrative, and healthcare services makes this exchange impossible, because the total autonomy of each district is protected.

1.7 Informal Settlements in Dar es Salaam. Taxonomy and History

Dar es Salaam's urban fabric is classified into four typological categories [11] in relation to the time in which they developed.

Old planned areas: urban areas built before 1970, including the city center, Oysterbay, Kariakoo, Upanga, Kinondoni, Ilala, Magomeni, Temeke, the industrial plants of Chang'ombe, and the area near Nyerere Road. These are all low-density residential areas with good urban services as they were reserved for European colonists. The city center and the area of Upanga present a medium level of population density with institutional, administrative, and commercial functions. Other areas, such as the Kariakoo district, display lower density with commercial and residential functions, retracing the boundaries of the original colonial settlement. Services and infrastructures are well distributed in this context, however they are inadequate in relation to number of inhabitants. The same situation can be detected in high-density districts, such as Magomeni and Kinondoni, originally devoted to the African population during the colonial era, with typical Swahili houses, rectangular buildings, regularly allocated within the plots, where sanitary conditions are quite problematic.

New, planned areas: developed after 1970, including, Kijitonyama, Mwenge, Sinza, Mikocheni, Mbezi Beach, and Tabata. These residential areas are characterized by a variable range of densities of dwellings (where this indicator is lower, quality-of-life standards are higher) and a good distribution of services and infrastructures.

Older, informal settlements: developed before 1980 such as Keko, Buguruni, Msasani, Mwananyamala, Hanna Nassif, Manzese, Mtoni, and Tandika. Again these districts are primary residential with an irregular distribution of buildings. Roads and routes are narrow, often located in open spaces in between dwellings where informal commercial activities can take place, facilitating pedestrian accessibility to the area. The quality of infrastructures and services in this context is low, and, in certain areas, they are completely absent.

New, unplanned areas: developed in the last 30 years such as, Ubungo Kibangu, Savei, Kimara, Mabibo, Goba, Boko, Bunju, Mbagala, and Ukonga. Many of the emerging informal settlements in Dar es Salaam fit into this category, located on the periphery of the city along main formal roads (Bagamoyo Road, Morogoro Road, and Pugu Road). These areas possess low-building density with no primary services or infrastructures. Unbuilt areas are exploited for agricultural and farming purposes. The quality of the constructions varies from zone to zone in relation to the wealth of the landowner.

Uncontrolled demographic growth and the consequent overcrowding of irregular settlements produce a high demand for formal services (electricity, water, etc.), decreasing their availability and efficient supply for the rest of the population. People are continuously exposed to sanitary, economic, and environmental risks, due to the lack or the unsuitability of water and sewage systems, as well as to pollution due to the inefficiency of the waste-management system. Even in more formal areas, the availability of such services is low and scattered. Major administrative financial resources are based on property taxes, urban services bills, local communities contributions and donations, governmental subsides, and excise duties [12]. In most cases, fiscal pressure on low-income inhabitants forces them to move to non-authorized irregular settlements. Dar es Salaam's housing emergency requires the construction of low-cost dwellings, basic services for the population, as well as employment for the occupants of such communities. Recent political changes in urban governance should foster participation and collaboration between local and national authorities in shared actions of territorial strategic planning. Even if the local government strives to promote forms of collaborative urban planning, formalizing property rights in informal areas in order to reduce the number of squatter areas, the lack of economic capacity to provide services makes these efforts unsatisfactory [12]. Still extremely relevant is the development of economic activities of subsistence such as agriculture and farming for a sustainable transformation of the territory. Such a planning approach arises from the need for technical implementation and optimization of the productive potential of these activities [13]. In this scenario, architecture still represents an irreplaceable resource to shape a new urban condition to formally establish the emerging integration between city and countryside as many informal settlements already try to suggest.

Analyzing the morphological development of many informal districts, it is possible to detect a set of common growth dynamics (Fig. 1.4). John Modestus

Fig. 1.4 Aerial view of the informal district of Yombo Vituka, object of this study (Google Maps)

Lupala's studies on the city's typological taxonomy [14], as demonstrated by the case of Msasani Makangira, indicates the evolution of a univocal pattern of aggregation behind the spontaneous development of these urban enclaves. Even if this model is repeated, without relevant mutations among all districts, the geographic conformation of the territory often implies a revision of the application of such rules in relation to location, exposure, and climatic conditions. Lupala presents three main stages of configuration: the first phase regards the occupation of the land, the second the consolidation of the original dwellings, the third phase outlines the densification and compaction of the urban cell. The first phase describes an ordinary mode of land appropriation in the case of peripheral grounds, for which informal settlements develop from small rural dwellings built on these areas (farms, stables, and huts) to defend and to cultivate the soil according to the presence of water and trees. If water allows the blooming of the crops, trees naturally compact the soil, and preserve it from crumbling during the rainy seasons, especially in proximity to rivers.

The spontaneous occupation of land for agricultural purposes is then directed by the presence of tall forest trees, from which it is possible to obtain edible fruits (mangos, coconuts, oranges, bananas). They also guarantee shading during the hottest months of the year. In the Tanzanian tradition, trees represent the main place of aggregation for the community. Around them the rural village is often built. The typical village is an irregular aggregation of dwellings and animal shelters surrounding an inner courtyard (a communal space for the use of all the families who live in the dwellings). In the middle of the courtyard, there are spontaneous or cultivated trees. Finally, in order to avoid flooding or landslides during the rainy seasons, a consistent distance can be detected between dwellings, grounds, and rivers. The phase of the consolidation of the settlement refers to the possibility of selling the crops to customers that are not part of the clan that owns the land. When the fields become very productive, the surplus of goods coming from agricultural activities can be sold. This case happens informally along the main routes and near the borders dividing each parcel of land or along streams and rivers where the soil is dry and resilient.

A community of farmers becomes a commercial society, shaping the spaces destined to sell goods, like those areas located along the narrow pedestrian paths that once were used to divide the rural properties. Next to these commercial facilities, all kind of services and cultural activities can take place in the developing rural-urban village. They take the architectural form of workshops and ateliers. Workshops are single-story, prismatic constructions with one face completely open to the street and with a closed back wall. Ateliers are made of light materials in order to speed up the building process, normally involving wood or metal sheets. Even if the appearance of these buildings may suggest that they are temporary, in reality they are strongly rooted in the urban fabric, underlining the path of public routes and open spaces, while protecting the private dwellings at the back with this system. At times, ateliers and workshops constitute a thick curtain between private and public areas, hiding clusters of private houses behind their walls.

1.8   Accessibility and Orientation: Emerging Suburban Issues

The process of densification of these settling structures represents perhaps the most complex and, at the same time, unreasonable building dynamic related to the growth of informal settlements in Dar es Salaam (Fig. 1.5). This is a process of infill between consolidated elements of the urban landscape, saturating the available space without considering minimum conditions of openness and exposure in order to make the places livable. The densification is caused by two main reasons: the need to obtain maximum profit from the fragmentation and the sale of rural properties (which have become unproductive as now they are completely urbanized), and the lack of other available grounds with similar geographic and hydrologic characteristics. People who are poor, or those new families that are less productive for a clan, settle in between plots and so filling all the possible gaps with additional constructions. These houses, in comparison to the older rural clusters, do not include an open courtyard, nor do they contains open areas for the execution of ordinary domestic activities. Sometimes the inhabitants of these buildings make an agreement with the landowner to rent the ground upon which they then constructed these temporary structures.

The process of densification constitutes the most relevant issue concerning the inhabitants of consolidated spontaneous settlements, especially because, in general, these minimum dwellings are built on the limits between one clan and another, one property and another, enlarging the problem of the so-called "inner backyards". The compaction and saturation of the space between houses recreates a series of inaccessible voids within the clusters. Such voids are preserved in order to offer the necessary light exposure to the houses and their windows, though they cannot be

Other urban Types / 2002
Informal settlements / 2002
Formal settlements / 2002
Other urban Types / 1992
Informal settlements / 1992
Formal settlements / 1992
Other urban Types / 1982
Informal settlements / 1982
Formal settlements / 1982
Main streets
Railways
Rivers
Sea

Fig. 1.5 Process of urban densification in Dar es Salaam from 1982 to 2002 [15]

accessed by the inhabitants. Often these voids are filled with garbage and they become unsafe as nobody is able to reach them without passing through someone else's property.

References

1. Lynch, K. (1964). *The image of the city* (pp. 1–2). Cambridge, USA: MIT Press.
2. Vidal de La Blanche, P. (1922). *Principes de géographie humaine* (p. 231). Paris: Colin.
3. Donatoni, F. (1970). *Questo*. Milan: Adelphi.
4. Farinelli, F. (2003). *Geografia* (p. 134). Torino: Einaudi.
5. Farinelli, F. (2003). *Geografia* (p. 176). Torino: Einaudi.
6. Ortiz, P. (2013). *The art of shaping metropolis*. Boston: McGraw-Hill Education.
7. Kombe, W. J. (1995). *Formal and Informal Land Management in Tanzania, the Case of Dares Salaam*. SPRING Research Series No. 13, Dortmund.
8. UNDP (United Nations Development Programme). (2007). *Lo sviluppo umano*, rapporto 2007/2008. *Resistere al cambiamento climatico* (titolo originale: *Human Development Report 2007/2008. Fighting climate change: human solidarity in a divided world*, Rosenberg & Sellier, Torino, Italia.
9. UN-HABITAT (The United Nations Human Settlements Programme), UNEP (United Nations Environment Programme). (2004). *The sustainable Dar es Salaam Project 1992–2003. From Urban Environment Priority to Up-scaling Strategies City-wide*. UNION, Publishing Services Section, Nairobi, Kenya.
10. World Bank. (1999). *Tanzania-Dar es Salaam Water Supply and Sanitation Project*. Washington DC, USA: World Bank.
11. Kironde, J. M. (1994). *The Evolution of Land Use Structure of Dar es Salaam 1890–1990. A Study in the Effects of Land Policy*. Ph.D. thesis, University of Nairobi, Nairobi, Kenya.
12. UN-HABITAT (The United Nations Human Settlements Programme) (2). (2009). *Cities and Climate Change Initiative, Launch and Conference Report*, Oslo, Norway, 17/03/2009, UNION, Publishing Services Section, Nairobi, Kenya.
13. Sawio, C. J. (2008). Perception and conceptualization of urban environmental change: Dar es Salaam City. *Geographical Journal, 174*(2), 149–175.
14. Cfr. Kyessi, A. G. (2002). *Community Participation in Urban Infrastructure Provision. Servicing Informal Settlements in Dar es Salaam*, Tanzania. Ph.D. thesis, University of Dortmund, Germany.
15. Lupala, J. M. (2002). *Urban Types in Rapidly Urbanizing Cities: A Typological Approach in the Analysis of Urban Types in Dar es Salaam*. Ph.D. Thesis, Department of Infrastructure, Division of Urban Studies, Royal Institute of Technology, Stockholm, Sweden.

Chapter 2
From Territorial Surveys to Mental Mapping. The Recognition of the Anthropological Image

Abstract In this chapter, the notion of anthropological image is applied to the understanding of the process of formation of informal settlements in prehistoric and historic sub-Saharan Africa, especially in the geographic region of Tanzania and Dar es Salaam. The concept of anthropological image emerges in this work from a semiotic reflection on Claude Levi-Strauss' investigations regarding the constitution of social structures in native communities. Levi-Strauss' creative and abductive approach to the construction of meaning in primary formations and his interest in rituals and their performance provides a methodology we can employ in the close readings of spatial configurations. Inhabiting units of nomadic and settled communities are analyzed in their parts and the way those parts are interconnected (Bantu, Tswanas, and Swahili). The role of musical (Muziki wa Dansi) and pictorial artistic practices (Tingatinga, batik) is also taken into account to offer a wide range of compositional systems to compare and distinguish recurring configurational practices.

2.1 The Notion of Anthropological Image

An anthropological image is a spatial scheme generated by the trajectories a community activates in order to inhabit a place according to common necessities and attitudes. Generally, an anthropological image is the result of an instinctive process of artistic transfiguration in relation to shared values and beliefs emerging from propitiatory rituals of appropriation of the land. Quite often, in complex and discontinuous conditions of urban growth, it is not always easy to detect the anthropological image that generated that settlement.

The inability to recognize the anthropological image of a settlement is inversely proportional to the degree of livability of a place, as the atavistic features that characterize the praxis of organization of the built environment remain hidden behind a set of superimposed elements introduced by rationalistic and exclusive projects. Anthropological images provide the appropriate spatial configuration for a community to acknowledge a place as its own settlement, suggesting rhythms and gestures for the topographic dance of inhabitation of an environment. Moreover, an artistic transfiguration constitutes a rendering of the real from the point of view of those who depict it. An anthropological image is a tool for the characterization of the place that outlines viable routes of navigation for the communities.

The elements that constitute the anthropological scheme of the settlement are to be researched among the artistic forms produced by an urban culture, narrative, musical, agricultural, or architectural. According to Claude Lévi-Strauss, the anthropological image of a place "...requires that we look at a population or a group of neighboring populations with the highest degree of practicality for what concerns the construction of the settlement [...] Exploiting a small number of myths taken from the native societies we are using for the research, we will experience something with a universal validity, as we expect from this to prove the existence of a logic of the sensible qualities of development that will manifest its rules of formation" [1].

A myth is a cultural expression, as ethnomusicologist Ernesto de Martino would say, that we can study on a rational level to understand the economic dynamics from which it was generated: "When the pain, with its polarities of pleasure and sorrow, and with its composite reactions, are framed within a rational perspective of production of goods according to a set of rules of action, chosen without restrictions and historically interchangeable, Vitality is resolved in Economy, and the human cultures emerges" [2]. The myth is the product of a natural economy that transforms actions into a ritual, or an expressive agreement. "The acknowledgement of the existence of an expressive convention—codified words and gestures, that can be transferred to others, that can be suspended or rejected, at times for individual choices, at times according to social obligations that imply hierarchies and times of intervention—all these matters can be explored only deepening the structure of the ritual discourse" [3]. The architectural ritual is manifested in the recurrent spatial behavior the inhabitants of a place enact when they move across a site. Learning the performative logics that support the construction of a built environment facilitates the investigation of spatial forms revealing the movements of the rituals they disclose.

2.2 Prehistorical Settlements

Urban farming is the most common production within the informal districts of Dar es Salaam, something rooted within an agricultural tradition developed in rural settlements from the countryside. In this geographical area, the South African settlement of Molokawe [4], from the ethnic group of the Tswanas—originally Bantu—constitutes a relevant archeological site to study the development of pre-historical rural villages. Even if the Bantus were originally from the north of Africa, due to several migrations, Bantu settlements were found in various parts of the continent, including East Africa (in the Tanzanian area) and South Africa. The Molokwane settlement contains circular settling units organized into more complex conglomerates. The unit is a small dwelling of few square meters. It is called

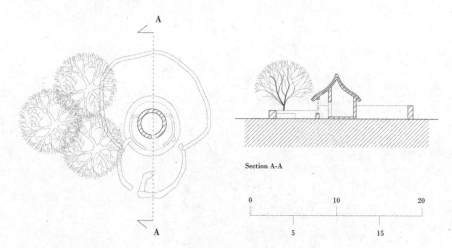

Fig. 2.1 The village of Molokwane, South Africa: morphology of the bilobial house

"bilobial" (Fig. 2.1) as it stands at the encounter of two walled rounded courtyards, one in front of the access to the house and the other one at the back. The courtyard at the forefront of the bilobial house was the most public part of the dwelling, from which it is possible to access another smaller open space before entering the inner rooms. At the back of the house, the other open courtyard was conceived for productive activities such as harvesting and the treatments of goods coming from agriculture and pastoralism.

In general, all houses are built in the vicinity of trees in order to provide shade during the day. Bilobial houses for single users are organized along a circle reserved for the male individuals of the community. The group of houses belonging to a family or a clan are called Kgoro (Fig. 2.2). In this configuration, women normally occupy the external area that surrounds the houses reserved for the men, while, in between each house and the inner enclave, a little open area is left for the treatment of the agricultural products. In this area, animals for the production of milk and meat are also gathered. They were located here to protect them from robberies and to avoid their escape. The settlement presents also another unit in between the size of the house and the size of the Kgoro. This unit is called Kraal (Fig. 2.3), in which single dwellings are assembled into groups of three units, each with an unbuilt area to guarantee access to the houses.

The combination of several Kgoro generates a type of settlement called Motse (Fig. 2.4), the sum of multiple circular nuclei, one next to the other, apparently without any transversal connection to link the dwellings. These settling forms grow together with the growth of the number of individuals within the families that inhabit them. On one hand, the settlement should enlarge its boundaries in order to allocate all the family members according to an eccentric vector, while, on the other hand, it encloses these people within an enclave that protects them from external forces.

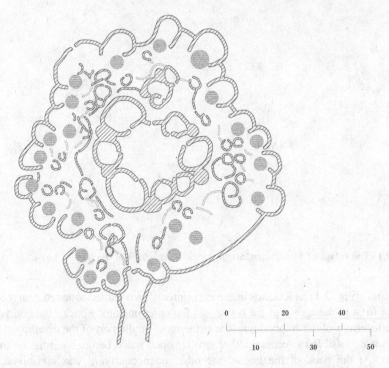

Fig. 2.2 The village of Molokwane, South Africa: the settlement of the clan called Kgoro

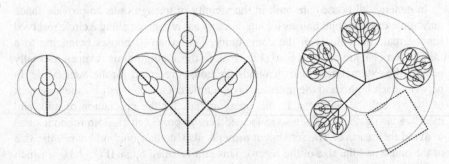

Fig. 2.3 The village of Molokwane, South Africa: configuration of multiple Kgoros

The area covered by each Kgoro is about 63 m^2. The articulation of multiple Kgoros follows curvilinear shapes in relation to the topographic conformation of the land on which they are built, not necessarily in closed circles. From a territorial view, the Motse is organized along major routes where all the accesses to the houses are located. At the scale of the Kgoro, this community seems to accept a circular enclosed conformation, while the Motse is effected by the vectorial force of the linear route (Fig. 2.5). The geometric identity of the nucleus is preserved,

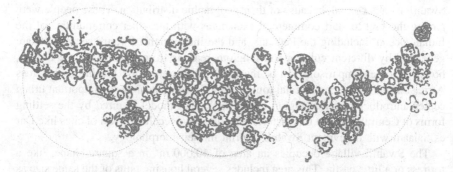

Fig. 2.4 The village of Molokwane, South Africa: multiple Kgoros aggregate in a linear figure

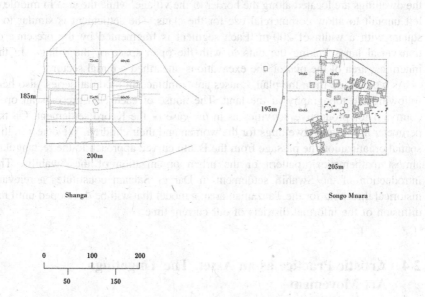

Fig. 2.5 The Swahili village of Shanga (*left*) and Songo Mnara (*right*)

although, potentially, the presence of other Kgoros and Kraals not directly connected to the main routes could generate more congested settling forms during the most advanced phases of their development.

2.3 The Configuration of the Swahili Settlement

The Swahili settlement instead is constituted of very different elements and forms. If the Bantus and the Twanas have nomadic origins, the culture of the Swahilis is sedentary, located in the east of Africa, in the region of Dar es Salaam. In Arabic,

Swahili means coastal, because of their geographic distribution. These people were among the first to start commercial exchanges with the other communities of the Indian Ocean, including the Persians, and the Indonesians. Swahili settlements are generated by different cultural and religious reasons from the ones of the Bantus, however some comparisons can be made. The diffusion of the Muslim faith across Africa through these commercial routes, made Dar es Salaam as an important urban center, introducing in this region a new settlement model inspired by the settling forms of Central Asia. At this historical moment, we recall the birth of cites like Dar es Salaam with an evident Swahili imprint in their morphology.

The Swahili village occupies an area of 40,000 m^2 in a square shape, like a fortress or a little castle. This area includes several housing units of the same size as the Kgoro settlement. In the archeological remains of Shanga and Songo Mnara, all the dwellings are located along the border of the village, while the area in middle is left unbuilt to allow commercial use for the clans. The settlement is similar to a square with a width of 200 m. Each segment is fragmented by the presence of transversal lanes linking the outside with the open space of the village. In the innermost area of the piazza, the excavations unearthed a burial sector.

As we can see from the plan, houses are combined in modular series, also here following the topography of the land. The house of each clan presents an open courtyard for productive activities as in the case of the Kgoro settlement. On the perimeter are located dwellings for the women and their children. We detect in this spatial organization the passage from the Bantu curved approach to the rectangular, almost mathematical, pattern of the urban organization of the Swahilis. The introduction of the Swahili settlement in Dar es Salaam constitutes a relevant historical threshold for the Tanzanian area, a model that will be developed until the diffusion of the informal districts of our current time.

2.4   Artistic Practice as an Asset. The Tingatinga Art Movement

Following the approach of Ernesto De Martino, in order to understand the character of Tanzanian cities, we combine the study of the morphological origins of their settlements with a broader investigation of artistic skills emerging from the cultural praxis of a community. We suggest here that, if artistic tendencies can be studied to explore rituals and anthropological uses, this should be applied to architecture, too, especially for what concerns the formation of the anthropological image of a city. In the case of Dar es Salaam, we refer to one of the most prominent artistic movements in Tanzania started around 1970, called Tingatinga [5], after the name of its inventor Edward Tingatinga. Tingatinga's works introduce us to an interesting research in painting, using recycled materials like pieces of wood and paint for bicycles. Bright colors and naturalistic topics are the main characteristics of these paintings, something particularly appreciated by tourists visiting Dar es Salaam in recent years. The case of Tingatinga can be considered as the beginning of a proper school of painting, as some of the pupils of Edward Tingatinga wanted after his death, in order to preserve and develop this style. When Tingatinga passed away, his activity called Tingatinga Partnership became the Tingatinga Arts Cooperative Society around 1990, following the artistic path of its initiator.

For the first ten years of existence of the Tingatinga movement, subjects were exclusively taken from the world of animals, with its patterns, its shapes, and its colors, which are particularly vivid in the Tanzanian environment. Tingatinga works of art exaggerate all these elements, translating them into an expressionist vision of nature. We could almost describe this art world as "naïve". Looking at these works, we can see an interest in what is enigmatic and complex, the result of a specific aesthetical intention emerged from a deep morphological research. Natural figures are composed and decomposed according to an articulated geometric order.

It is interesting to note that the second generation of painters of the movement actually introduced some relevant innovation in the Tingatinga style: after the death of the master, during the 90s, his pupils adopted a new set of subjects for their paintings. Since then, even the city and its transformations have encountered the attention of the artists, revealing some important social changes of the Tanzanian metropolis. Perspective was introduced as a technical tool for representation, passing from a natural chromatic environment to a regulated art space with a geometric organization. Such a tendency coincides with the meeting of some parts of the population with art dealers coming from Europe, the United States, and Asia. Tingatinga paintings are generally created for tourists on small formats in order to be transportable on a plane. The little canvases are realized on masonite with layers of non-dilute varnish which provides a typically lucid texture to the drawing, with the boundaries of the figures in relief. More recently, Tingatinga naturalistic subjects have been applied to the decoration of pottery and other objects too.

2.5   The Batik Painting Technique

Very popular in the same geographic area is also the batik painting technique. Batik represents both a pictorial subject and a writing technique, as emphasized by the name itself, which means "drop writing". The term has Indonesian origins: "amba" means writing, while "titik" means point or drop, and the action of the artist is called "membatik". The batik is a technique for the wax-resistant dyeing of fabrics. Drops of wax or grease applied to a cloth prevent it from absorbing the color. When the wax is removed, the canvas assumes the typical striping that confers varied nuances to the drawing. The batik—very popular in India and on other Continents—reproduces in Africa recurring subjects such as the stroll of the women loaded with goatskins toward the pit or carrying out domestic work, some of them with their children on their shoulders. This is a very frequent image in rural Africa, where women are required to take care of all these labors. In Tanzanian batiks the human figures are strongly stylized on a multi-colored, almost surreal, background obtained using the drop-writing technique. We could say that this technique implies the construction of the image in negative, in relation to the picture we intend to create. With the batik technique, the composition is somehow inverted, as what is designed first is the configuration of the drops from which the final outline will emerge after their removal.

2.6   Musical Composition. The Muziki Wa Dansi

The city of Dar es Salaam was also particularly renowned for the musical development of the country since the 1930, when across its clubs the Muziki Wa Dansi spread the oldest musical genre in Tanzania. This genre descends from Soukous music, originally from the Congo and characterized by a rumba rhythm mixed with an international swing style. The Muziki Wa Dansi is a dance in four that includes elements of jazz and the swing tradition. A part from the typical percussive rhythm—a characteristic of all African music—vocal interventions alternate with instrumental ones. Also, non-percussive instruments used for this music are guitars and winds coming from the American and the European traditions. Vocal interventions could be solos or include more voices. Choral interventions are typical of this music, linking this sound to older tribal songs where voices always appear in ensemble, developing the melody homophonically, though according to local harmonic rules.

This condition provides a local quality to a music type that could easily be considered as international. Muziki Wa Dansi is the result of a mixture of various cultural influences in an heterogeneous, inclusive way. This music renovates the sound of various tribal songs toward a global and modern development. As proved by several cultures in history, dance motives are among the most resilient music types for a community, because, through sounds and gestures, people find a way to connect their bodies to the topographic features of a land [6]. Other dance styles in Tanzania that have developed more recently are the Taarab, coming from Zanzibar, and the Bongo Flava, an African interpretation of the international hip-hop genres. Both music types are dances with an incisive percussive rhythm and a simpler organization of the vocal parts.

References

1. Lévi-Strauss, C. (1964), *Il crudo e il cotto*, Bollati Boringhieri, Milano (2000), p. 13.
2. De Martino, E. (1958). Morte e pianto rituale nel mondo antico: dal lamento pagano al pianto di Maria, Bollati Boringhieri, Torino (2000), p. 18.
3. De Martino, E. (1958). Morte e pianto rituale nel mondo antico: dal lamento pagano al pianto di Maria, Bollati Boringhieri, Torino (2000), p. XXXVII.

4. Hall, S. (1995). Review of Pistorius, J. C. C., 1992, Molokwane: An Iron Age Bakwena Village, Perskor, Johannesburg. *South African Archaeological Bulletin, 50,* 88–89.
5. Banks, P. (2010). *Represent: Art and identity among the black upper-middle class.* London: Routledge.
6. Warburg, A. (1988). *A lecture on serpent ritual.* London: The Warburg Institute.

Chapter 3
Urban Samples. The Reconstruction of the Settlement Model

Abstract The agogic map of Dar es Salaam is built upon an urban quadrant studied in its geographic and anthropological characteristics. The choice of the quadrant is influenced by the intention to work within a context of informality in proximity to other more structured parts of the city. The chosen urban sample embraces a large variety of informal settling features, and it is close to major infrastructural elements. The presence of these elements constitutes a reference for the modern development of informal enclaves, although a sustainable exchange between all parts requires a revision of the current system of accessibility and spatial orientation. Combining the elements of the anthropological image of the settlement detected in the previous chapter, with Kevin Lynch's five urban elements of the image of the city, a mental map of the informal district is reconstructed and discussed as a possible structure for the development of an agogic map of Dar es Salaam.

3.1 Sustainable Settlement Models in Urban-Rural Metropolitan Scenarios

This research investigates the implementation of new settlement types that are generic and flexible as well as sufficiently structured in order envisage a feasible scenario of urban growth for areas that are irregularly built and occupied. The current perspective of integration between urban configurations and rural forms of land colonization presents three models of development that could be applied to the case of Dar es Salaam: the Desakota model developed by Terry McGee, the Mega-block model introduced by Grahame Shane, and the BUD model theorized by Pedro Ortiz.

The Desakota model emerges from urban geographer and social scientist Terry McGee's studies about Southeast Asia and in general about those geographic areas that could be defined with the obsolete notion of Third World, including Africa. The model proposes the creation of urban-villages as new metropolitan units

(*Desa* stands for village, while *Kota* means city), combining rural territories and urban settlements [1]. This model promotes the construction of a technological, highly-proficient, rural landscape enabling new modes of high-speed motion within a large region, while preserving all the necessary conditions to enhance slow-mobility. This is a typical condition in rural villages and farms. McGee describes this model as a "trans-active environment", in which the countryside and the city develop in parallel, while exchanging assets coming from the exploitation of both landscapes. A large-scale infrastructure made of streets, canals, and water control systems guarantees the efficiency of the entire urban region, while the rural architectural features of historical settlements are preserved. The Desakota model applied to the context of Dar es Salaam could represent an effective approach to the transformation of informal areas that include a significant number of rural activities.

The Mega-block model constructed by urban theorist Grahame Shane takes advantage of many features implied by Terry McGee's Desakota model such as the urban-rural character of the metropolitan units and the productiveness of the landscape for the economic sustainability of the settlement. However, this model makes a point in regards to the size of the interventions. Shane's Mega-block measures 1 k × 1 k, with a large variety of functions and forms within a sub- stantially rural fabric. This fabric is more complex in forms and functions in comparison to the Desakota model in the way it combines and overlaps a large variety of urban uses. Each tone of the landscape in this model is like an exchanging patch, a patch-dynamic [2], where functional features are mixed and ambiguous. What is mono-tonal does not activate the urban life of the Mega-block, therefore all the layers are intertwined. The Mega-block superimposes "grey" ele- ments for the movements of the inhabitants on a "green" productive territory for a proactive exchange between the two systems.

The BUD model (Balanced Urban Unit) [3] insists on an even larger size of intervention, covering an area of 5 k × 5 k. It potentially includes multiple Mega-blocks. Urban planner Pedro Ortiz defines the BUD model as an urban archetype to apply each time differently according to the topographic condition of the territory encountered. The BUD model is an urban unit in which a city demonstrates its accomplishment, i.e., its development at the scale of a metropolis. This model is the result of various investigation on multiple case studies around the world for different metropolises. The 5 × 5 k block incorporates the landscape as a design variable, as a tool for the environmental temperament of the city. The system exploits a commuter train to connect all the locations of the urban region. In the middle of the system, we can find a train station with direct access to the historical nucleus of the settlement, while, along major regional routes, we recognize the so-called "great regional equipment", mostly industries and commercial plants for the sustenance of the metropolis. The model is assessed comparing thirty case studies and numerous master plans conceived for such cities. The measurements of the BUD model do not seem to recall any specific anthropological rule, nor image, however the empiric confrontation of different metropolises highlights the relevance of the topography concerning the localization of the model within a real context.

3.2 An Exemplary Metropolitan Section: The Informal District of Yombo Vituka in Dar es Salaam

The urban sample we have chosen in Dar es Salaam is the informal district of Yombo Vituka (Fig. 3.1), less than 4 k from the city harbor, South of Pugu Road, half way between the old town center and the airport. Pugu Road defines the northern boundary of Yombo, while the informal district of Tandika (Old Informal Areas) and the formal settlement of Temeke (Old Planned Ares) represent the western and the eastern boarders of this area. Yomobo Vituka is included within a quadrant of about 5 × 5 k, containing a wide variety of functions and morphologies, regular and irregular, a train station for commercial exchanges between the city and the region, and a large unbuilt strip occupied by the presence of the railroad between Yombo and Tandika. Finally, the project presented in this investigation requires a focus of about 1 × 1 k, aiming to validate the introduction of innovative intra-scalar settling units within a BUD model applied to Dar es Salaam. This smaller quadrant is included between the station, the border of Tandika, Makarangawe, and the industrial area of Kipawa.

Although the area is mostly flat, the morphology of this territory shows the presence of numerous natural water basins derived from the River Kinziga. Most of them are seasonal streams that follow the Kinziga until it encounters the sea near the harbor of Dar es Salaam. This condition suggests that the slope of the soil declines towards the southeast. Infrastructures, in fact, are located along Pugu Road on dry, resilient land. Unbuilt areas present florid spots of spontaneous vegetation due to the humid quality of the soil on these sites. Next to these spots, we detect also the presence of small rural plots along watercourses and canals. Along Pugu Road, next to the railways, there are also numerous industrial plants, including, among others, some aluminum refineries, paper makers, furniture builders, and

Fig. 3.1 Dar es Salaam, Tanzania. The informal district of Yombo Vituka

logistics companies. Such activities are directly related to the presence of water and canals to support the processes of production and transformation of the goods.

The unplanned areas emerging in between the outlines of the basins and the main infrastructures are occupied and often saturated by the presence of spontaneous settlements and illegal dwellings. The proximity with any water feature is an asset for those who settle on these sites, however, during the rainy seasons, these areas are off limits because the soil becomes very unstable and wet. Informal settlements seem to be very compact urban entities enclosed within the lines of railways and motorways. Yombo Vituka is rooted around some industries and ateliers, while it seems to settle away from water features and the major infrastructural system. Moreover, streams and natural canals influence the morphology of Yombo, employing dry lands as the more resilient geographical environment for the urban development.

On the other hand, the structured city interprets the "informal blockage" as a threat to the future expansion of formal districts due to the impossibility to include informal settlements within a more inclusive metropolitan plan. Informal settlements in this sector present a special relationship between the countryside and somehow the anthropological modes of cultivation and transformation of the land, introducing residential functions and forms typical of the rural communities of Tanzania, divided into clusters of clans and families. In the western part of the quadrant, a more complex morphological case emerges where an exchange happens between the residential blocks of Makarangawe and the main infrastructural routes. At this point, a process of urban movement between two different morphologies can be detected.

Here some urban elements, which in usual metropolitan conditions should be able to create centralities and exchanging patterns in reality, are inhibited by the inability to establish long-term effects on the transformation of the district. Sites like the railway stations of Magakawe or Yombo, or even the industrial plants along Pugu Road, are more like barriers instead of ecotional thresholds that structure the overall urban system. At the same time, at the small scale of the informal district, informal dwellings introduce a recurring constructive persistency that constitutes a reference for the navigation of the city. Such settlements establish an innovative compromise between urban and rural environments within the city-region.

3.3   The Mental Image of an Informal Settlement: Centralities, Enclaves, Landmarks, Paths, Densities

Within the urban sample of Yombo Vituka, we can detect topographic discontinuities passing from the ancient to the modern and from the rural to the urban. The reconstruction of the settling model reflects this condition. The construction of the map of the informal district of Yombo Vituka starts outlining the presence of major formal built elements. Concerning the informal features of the quadrant, the topography of the land is investigated, highlighting basins and orographic emergencies.

Following Kevin Lynch's procedure for the individuation of the urban image, paths, margins, districts, nodes, and references should be identified; however Lynch's approach is conceived for formal cities like Boston and New York. For this reason, as we are working on an informal sector, the parameters set by Lynch are assumed from the characteristics of the anthropological image of the rural village as the archetype for the development of informal districts within the city of Dar es Salaam. Looking at the Molokwane village, we acknowledge the centers of the settlement, boundaries for inclusion and defense, territorial landmarks, and vectors of navigation.

Centralities in informal settlements can be found within the circular suburban isles of the city fabric (Fig. 3.2). In between various units, there are areas of friction where other forms of centrality emerge. Those centralities constitute an attractive system at the scale of the district. They are localized at the meeting point between various paths, generally where a vector divides into two or more branches. For example, where the system of pedestrian paths along the river develops into more routes, this condition provokes an enlargement of the street section, recreating a triangular square. On the short basis of the triangle, we can normally find an atelier, a shop, or a service for the community.

Looking at the organization of the paths within the informal district of Yombo Vituka, there is also a longitudinal system of semi-public pedestrian paths. The continuity between each unit is avoided by the presence of a building or an atelier, as if the general flux of people is inhibited in moving along this network of routes. For this reason, we call this system "semi-public", referring in particular to the fact

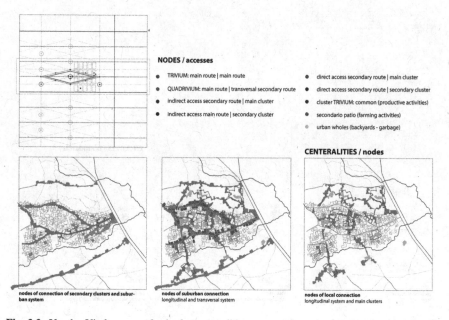

Fig. 3.2 Yombo Vituka: map of suburban centralities

that these streets are for the private use of the family or the clan that occupies each circular unit.

The morphology of the informal district presents a direct continuity between social forms of aggregation and the spatial configuration of the habitat. The clusters of dwellings are normally constituted by three to five prismatic single-story dwellings (20–40 m^2) organized around an open courtyard in the middle of the informal urban isle. These dwellings are not aggregated in compact structures, instead, they are free standing, allowing the formation of secondary paths and visual trajectories in between houses. This irregular configuration includes also a number of smaller private courtyard for farming activities, and it implies the shrinking of the thresholds in order to make the access more controlled and defensible. Courtyards in the center of the urban isle are wider and articulated, normally with a wide tree in the middle. The houses that surrounds this open space are the oldest buildings of the island, often belonging to the first family who occupied the land on that site. The relevance of this cluster is confirmed by the presence of wider accesses to the central courtyard, while other conformations of buildings within the same isle overlook smaller and less accessible courtyards. The access to the entire island is regulated by few thresholds in relation to the relationship that the inhabitants and the clans intend to establish with the external urban entities: the more heterogeneous the community is, the greater is the number of accesses.

Analyzing thresholds and boundaries of the spatial objects of the quadrant (Fig. 3.3), the morphology of the settlement shows that a building is always located at the end of a route [4]. Near rivers or water basins the shores (width 30–50 m) are

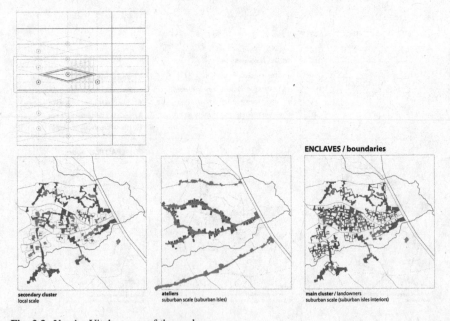

ENCLAVES / boundaries

secondary cluster
local scale

ateliers
suburban scale (suburban isles)

main cluster / landowners
suburban scale (suburban isles interiors)

Fig. 3.3 Yombo Vituka: map of the enclaves

left unbuilt for safety reasons. Each urban isle occupies an area of about 2000 m², a square with a side of 30–50 m. In order to understand the origin of this measure, it is necessary to study the configuration of the cluster that belongs to the owner of the land. In general, one single building constitutes the origin of the cluster, a rectangle of a single story, normally located next to a tree. The presence of trees provides shadow and cooling during the hot seasons, but it indicates also that there is water in the soil and therefore that the land can be treated for rural purposes. The trees becomes an indicator of potential urban development, as from one house the family stretches out to other buildings creating a courtyard around the tree. The tree is the point of aggregation for the family to enact all the productive activities, farming, cooking, weaving, and even entertaining. When the farm becomes part of the city due to the current dynamics of conurbation, clusters multiply into a group of constructions until all the property is filled with buildings.

In the context of an informal settlement, the urban block (or suburban isle) was once a farm. The outline of this entity is retraced by a set of ateliers along the main routes. The morphological studies on this region, collected by urban researcher John Modestus Lupala, recall the rural origin of these lands before the activation of the great fluxes of immigration in recent times, which necessitated the fast construction of emergency shelters and irregular houses everywhere [5]. However, as proved by the research on Yombo Vituka proposed in this book, this development follows in most cases the rural division of the "fields". This means that most of the routes that were used to indicate a separation between rural properties are now transformed into pedestrian, suburban paths. The routes dividing farms and fields were also the most appropriate location to sell rural products directly to travelers or people passing by. In this location, products could be collected and then carried away to local markets or other villages. For this reason, together with the increase of migrations, little shelters and provisional ateliers were built along the routes, for the farmers to harvest and sell their products. In order to control effectively the rural property, the farmer's house was located in the center of the land, while ateliers constitute a link between the most private part of the property with its public thresholds.

Dynamics of compaction of rural properties (Fig. 3.4) confirm the central cluster as a resilient built element that structure the new spontaneous urban fabric. Around this point, until the boundaries of the property, the land is now filled with houses. The central cluster is recognizable because it always contains an imaginary line that divides the isle into two equal parts. The central cluster, once the fundamental element of the rural organization of the land, seems to interfere with the new urban configuration of the district: the cluster is inward-looking, like an enclave that has little relationship with the public routes surrounding the suburban isle. This condition provoked the formation of an autonomous north-south network of semi-public streets to interconnect all the clusters, thus avoiding public routes.

In the degenerative phase of compaction of the land with houses and clusters, a condition of ambiguity is created between the notion of front- and backyard. This situation in recent years has generated a very relevant issue of inaccessible backyards: narrow open spaces constricted between the perimeter walls of different

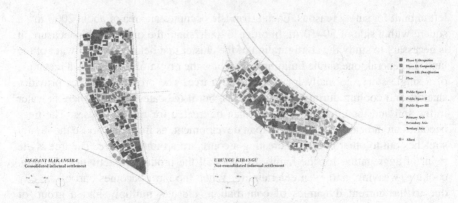

Fig. 3.4 Dar es Salaam. Analysis of the development of the informal settlements of Msasani Makangira (*left*) and Ubungu Kibangu (*right*) (Lupala 2002)

houses remained empty to guarantee air and light to the dwellings. These spaces are not reachable if not through private houses. These irregular spontaneous backyards are very important for the survival of the settlement as they provide natural ventilation during the hot seasons. However, they also represent an issue of health and sanitation because most of these plots are filled with garbage and used as private dumps.

A specific study on the spatial density of the settlement of Yombo Vituka focuses on the degree of separation between buildings. This parameter indicates that, the greater is the density of the polygons in which each building is included and the smaller is the area covered by the construction, then the greater the sense of division among all parts. This condition is directly related to the property of each structure. Bigger dwellings belong to larger family and clans, normally rooted in the land they inhabit, possibly inherited directly from the first colonists of the land. Smaller plots belong to immigrants and newer families. The highest degree of openness in this system is detected when the width of the street section varies between 20 and 50 m. The lowest level of accessibility of the spaces in between houses is between 0 and 5 m. Between 5 and 20 m of width of the street section, the analysis shows the presence of an urban environment whose accessibility is regulated by the land owner of by the members of the clan.

The functional differentiation of communal spaces within the informal district seems to be directly related to the formal articulation of unbuilt areas along the paths. Where the vector of a path changes direction to encounter a new branch, the meeting point between the two linear elements is a wider open space that can be catalogued and defined according to parameters such as distance from the water features and the slope. The buildings located along this system are very important to understand the way the spatial movement is configured in this context (Fig. 3.5).

Even if informal settlements are shifting and continuously changing due to their spontaneous transient nature, some elements are more resilient. They are landmarks kept on a longer term to mark the landscape. The openings in the compact line of

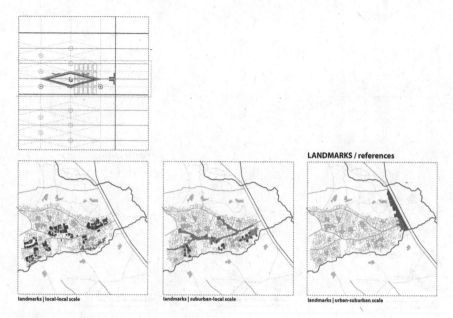

Fig. 3.5 Yombo Vituka: maps of the landmarks

ateliers along the borders of the urban isle can be interpreted as resilient architectural elements that mark the presence of the central cluster, the family of the first colonist who—in most cases—still inhabits the land. At the local scale, the presence of the central cluster is also the marker of the network of secondary pedestrian paths interconnecting all other clusters. Finally, at the scale of the city, the passage from formal sectors to informal districts is established by a much wider and evident threshold along which there are less built elements, where pedestrian routes unfold becoming paths into the landscape. This can be also detected along rivers and water basins. In certain cases, like the north connection of Yombo to Airport Road, a voided strip with a width of 30–50 m indicates a clear separation between the two environments. On one hand, the strip is still a porous line that allows people to enter the informal district on foot, while, on the other hand, this line defines a border that cannot be easily overcome (Fig. 3.6).

Finally, a topographic investigation of the conformation of the settlement provides an overview on the geographic distribution of all the elements of the anthropological image of this built environment. The topographic is employed in this study to test all the conjectures arising during the morphological analysis of each element. Topography should answer the final question: Why a recurring element is located in such place? Why would it assume such specific shape?

Zooming in on Yombo Vituka and looking at the topographic conformation of the place, it is possible to notice that all the main routes move in parallel with the direction of the river, marking the presence of several shores at different heights. In fact, the posture of each path follows in fact a contour curve of the soil. The

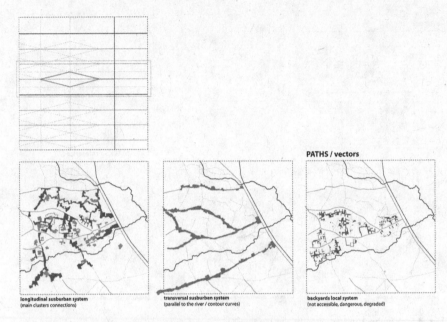

PATHS / vectors

longitudinal susburban system
(main clusters connections)

transversal susburban system
(parallel to the river / contour curves)

backyards local system
(not accessible, dangerous, degraded)

Fig. 3.6 Yombo Vituka, Dar es Salaam. Maps of pedestrian paths

network of the so-called semi-public routes is instead organized transversally in respect to the network of main connections. Where they encounter the main system, they outline the boundaries of urban cells, the suburban isles. Ideally dividing the area included between the main route closer to the river and the coastal line within five lines and four spaces, like a stave (Fig. 3.7), the houses are generally located within the spaces while, as we said before, the central cluster sits on the lines. Central clusters normally occupy an area that is dry and easy to defend, confirming their importance in the organization of the settlement.

Confronting spatial density and routes, the settlement presents another important relationship between the size of buildings and the width of streets. If we look at the parametric map of the densities, three thresholds of accessibility emerge. The

Fig. 3.7 Yombo Vituka, the Stave figure: the topographic configuration of the settlement

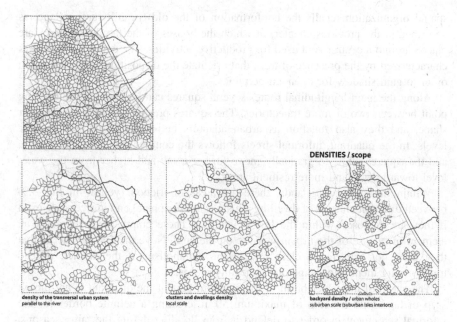

DENSITIES / scope

density of the transversal urban system
parallel to the river

clusters and dwellings density
local scale

backyard density / urban wholes
suburban scale (suburban isles interiors)

Fig. 3.8 Yombo Vituka, Dar es Salaam: map of building density

greater is the distance between two neighboring buildings, the bigger is the degree of public accessibility (Fig. 3.8). The three thresholds express the low density of main public routes, the medium density the paths of the urban isles, and the high density expresses the presence of inaccessible backyards, where the passage is obstructed and movements are unsafe and unhealthy. The thresholds individuate the degree of separation between spatial objects while defining the urban qualities of some vectors.

3.4 Anthropological Images and the Reconstruction of the Settlement Model: Courtyards, Squares, Backyards, Shores, Paths

The study of the anthropological image of the settlement of Yombo Vituka delineates a degenerated settling form, originating from the transformation of rural villages and farms into overcrowded semi-urban residential structures. Accepting the condition of the built environment of Yombo as a spontaneous self-sustained habitat, the elements proposed by Lynch applied to this case study have been identified as follows: courtyards, squares, backyards, shores, and paths.

Houses are mostly distributed around courtyards (blue in Fig. 3.9) located away from major streets and pedestrian routes in order to make them defensible. This spatial organization recalls the conformation of the old Swahili rural village, as described in the previous chapter, in which the houses of the family members are placed around a central void used for productive activities. The courtyard is always characterized by the presence of trees that orientate the disposition of the houses in order to gain shadow for open–air activities.

Along the main longitudinal routes several squares can be found at the meeting point between two or more trajectories. The squares present normally a triangular shape, and they also function as urban adaptors between different topographic levels: In the quadrant, informal streets follows the contour curves of the ground, and the squares allow to gradually move further away and higher from the water level toward dryer and more resilient areas.

Within the suburban islands, there are areas of friction (yellow in Fig. 3.9) between different clusters of dwellings that remain peripheral and inaccessible even if they are enclosed within the line of ateliers. These empty areas represent an element of release of the urban pressure on the land. Nevertheless, the fact is that they have not achieved a clear role in the metabolism of the district, instead transforming them into dangerous unhealthy places.

The shore of the water basin in the southern part of the quadrant is an unbuilt strip of land with a width of maximum 50 m, creating a definite border for the informal settlement in order to defend it from flooding during the rainy seasons. Shores in this areas can be up to 5 m high, and they constitute an off-limit line beyond which it is not possible to build.

Paths in Yombo Vituka are normally parallel to the watercourses, often following the orography of natural basins and hills on which the settlement is built. Paths should be interpreted as routes that connect houses to the resources necessary to sustain the community. Two main networks can be identified, a longitudinal one (east-west), parallel to the river, linking the urban isles to the rural area enclosed by two railway lines toward the east, and a transversal one, passing through all the urban isles and intertwining the main clusters of each isle to avoid using main routes (Fig. 3.10).

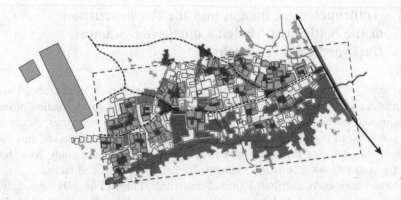

Fig. 3.9 Yombo Vituka, Dar es Salaam: map of the urban elements

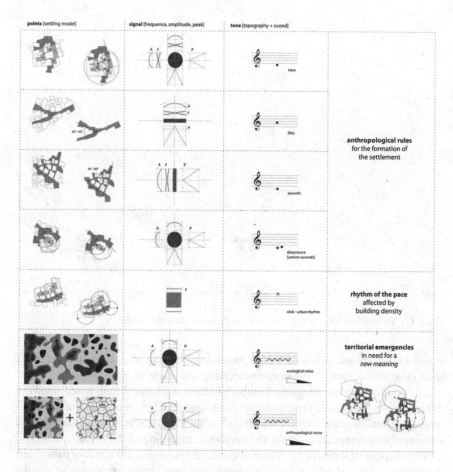

Fig. 3.10 Yombo Vituka: matrix indicating sound and spatial proportions in relation to the most prominent urban elements within the agogic map

The finding of the fundamental elements of the anthropological image of Yombo Vituka provides all the components we need to reconstruct the settling model of the community, an immanent structure that can be employed to systematize the construction of the informal district while envisaging the future urban development of a metropolis that is facing issues of uncontrolled demographic growth, excessive urban pressure, and infrastructural weakness. The model is a geometric system that is essential for a parametric understanding of the urban dynamics of formation and transformation.

We call the model described above "the Stave of Yombo", as the five contour curves of the shores are repeated symmetrically in the northern part of the quadrant and then similarly over the river in the southern part (Fig. 3.11). The four shores of the spaces between the lines represent the space available for construction. All the

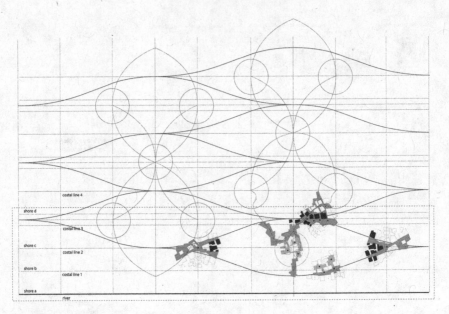

Fig. 3.11 Yombo Vituka: reconstruction of the generic settlement model

constructions are subordinated to the presence of original clusters at the center of what once was a rural compound, intercepting the shore lines.

The anthropological image of Yombo Vituka is a relational map based upon the distance or the degree of separation between spatial objects within the urban environment. Detecting the metric rules that regulate the configuration of the informal settlement constitutes an asset for the geometric modeling of the urban system.

The resilient structure of the informal district displays also a secondary palimpsest of smaller routes which perhaps represents the most transformational of all the elements of the settlement. These paths become inaccessible very easily following the usual dynamic of compaction of informal areas. The focus on the shifting elements that characterize the quadrant establishes the passage from a static morphological study of the built environment to a temporal consideration of the urban artifacts: the inclusion of the time among the variables of spatial definition enables one to outline the landscape in terms of spatial behavior and movements. This approach is suggested in this research as particularly appropriate to interpret informal settlements, especially in regard to their moving and quickly-changing conditions.

References

1. McGee, T. G. (1971). *The urbanization process in the third world*. London: Bell and Sons.
2. Shane, G. (2005). *Recombinant urbanism*. London: Wiley.
3. Ortiz, P. (2013). *The art of shaping the metropolis*. New York: McGraw Hill.

4. Farinelli, F. (2013). *Geografia*. Milano: Einaudi.
5. Lupala, J. M. (2002). *Urban Types in Rapidly Urbanizing Cities: A Typological Approach in the Analysis of Urban Types in Dar es Salaam*. Ph.D. Thesis, Department of Infrastructure, Division of Urban Studies, Royal Institute of Technology, Stockholm, Sweden.

Chapter 4
Agogic Maps. A Topography of Sound Signals for Spatial Orientation and Configuration

Abstract The concept of anthropological image, intertwined with Lynch's fundamental elements of the image of the city are exploited in this chapter to establish an initial itinerary of locations and distances for the construction of the agogic map of Yombo Vituka. Musical notions in spectralism, psychoacoustics, and ethnomusicology all contribute here to generate a creative cognitive scenario of sound signals to foster the experimental extent of the map. Environmental, social, and spatial conditions of the place are described through a series of sounds that orientate and make these locations memorable in order to increase the accessibility and the livability of informal suburban structures. The agogic of the place is spatialized through a digital prototype (Geoscore), and its extension is outlined through a comparative table, where the sound spectrum is analyzed according to environmental, social, and time variables. The space acquires a performative scope. The project becomes a chronographic rendition of spatial transfigurations.

4.1 The Notion of Agogic and Its Relevance to Space-Time Design

The term Agogic, as described by Hugo Riemann, is the practice of rhythm [1], a practice for which the extent of each measure—or the cognitive distance between the body and the space—varies. Such variation provokes in the body an expectation in regards to the measures perceived, as a matter of experience of a combination of space and time. Spatial configuration requires the reception of each spatial mark as spatial pulsations, as if each sign could create attraction and involvement within the inhabitant.

In the early Western musical tradition, an agogic motion converts a consistent rhythm in a fluctuating horizon, normally on the basis of a poetic text. As English anthropologist and writer Bruce Chatwin reminds us, often in certain cultures tunes and songs are preeminent mapping tools for their capacity to provide meaning in describing the relationship between our body and the topography of a place [2].

45

The agogic scoring of the space discloses a spatial behavior musically accepted by a community. Agogic maps indicate possible spatial habits within a site, setting to music the anthropological image that gave rise a settlement or a landscape. Agogic is the discipline that regulates the mental and physical pace of individuals moving through an unknown geography, employing musicality as an experimental tool for spatial appropriation. We could claim that agogic is a protocol to access the sensitive extent of the space, like an instrument of spatial and cognitive orientation. Agogic maps require the identification of rhythmical variables of spatial definition to determine the in-motion configuration of the space, identifying a tune that guides the body in the process of acceptance of the qualifications of livability of the place.

4.2   The Pace of Inhabitation. Applying Agogics to the Anthropological Image of the Place

The pace is the expression of material and immaterial aspects of a built environment. From the study of the rhythm of spatial movements, we can understand morphological dynamics that are difficult to interpret if we simply look at the static disposition of architectural objects. The case of Dar es Salaam and its informal district is particularly telling, as in this context it is difficult to distinguish accessible routes from dangerous or inaccessible ones at first sight.

On this matter, the words of Émile Jaques-Dalcroze, the founder of Eurhythmics, are very relevant to acceptance of agogics as a time-space discipline. His books, written during the 20th century, clarify the potential implication of agogics for the development of a new generation of mapping tools: "[Those] who progressively educate their own bodies to the rhythm and the dynamics of music, they will be more musical themselves, and they will be able to interpret spontaneously the intentions of the composer. Walking or dancing on Bach's fugues will not ever be a crime against the great genius [...] this corporal interpretation does not intend to accomplish the author's ideas, nor to substitute his means of expression with an arbitrary translation. This is an inner journey that replaces a fully intellectual analysis with the experiences and the sensations of the whole body [...] for the pupil highlighting the different voices, disassociating polyrhythms, fulfilling strettos or understanding the nuances of each dynamic will be something natural and intuitive. [All these elements] will be clear to him because he will have experimented them through his own body. This organic quality referred to compositional means contributes to develop the human instinct, fostering the establishment of a healthier social life" [3]. Dalcroze assigns to music a prominent role in involving the body within the expressive configuration of a space. The tune that shapes and orientate the construction of a settlement on a site is an expression of the anthropological image. This image establishes an original spatial structure upon which the language a community is articulated in order to settle in a certain context. It is like a *leitmotiv* of the artistic intake of the inhabitants of a place applied to the geographical materials of the place.

4.3 Geoscore. A Prototype for an Agogic Map in Dar es Salaam

An agogic map helps in identifying and displaying such themes, introducing an innovative way of communicating the expressivity of the space intertwining sound and digital mapping (http://www.youtube.com/watch?v=RF3ZzyKXsrQ). The map is an immersive tool for orientation that coordinates urban movements according to a screenplay of images and sounds along various geographic lines.

In the case of Dar es Salaam, the screenplay combines a geometric model of the informal settlement of Yombo Vituka—as outlined in the previous chapter—with two interfering layers. The objective of this operation is to experiment with the depth of the space through a digital tool that recreates a multi-dimensional experience. This experimental map marks all the resilient elements of the built environment and classifies their relationships with ecological and anthropological operators [4].

The ecological layer refers to environmental cyclical motions, while the anthropological layer traces the accessibility of the urban fabric according to values of spatial density. The settlement is analyzed focusing on its adaptations to values of soil humidity and temperature change in relation to the distance between objects. They define respectively the degree of resilience of the built environment and the degree of accessibility and livability of the place. Moving elements of the map correspond to dissonant sound signals, affecting the perception of safety and accessibility typical of more resilient, established spatial objects. This approach invites the user of the map to assign a meaning to those sensitive and weaker sites with an unstable urban identity without imposing external definitions.

4.4 The Green-Grey Series

The layer representing eco-systemic interactions between the environment and the settlement (Fig. 4.1) provides an interpretation of the range of "green" tones detected within the quadrant. Each tone discerns different conditions of bio-diversity, intended as a morphological and locative variety of the architectural elements of the district. The tone presents certain formal outlines emerging from the topographic characteristic of the site. The approach employed in this research intends to study the topography of the built environment of Yombo Vituka from a deterministic point of view, using a parametric technique of analysis while emphasizing the relevance of chromaticism as an instrument of spatial orientation.

A tonal approach to the geography of the place transforms the settlement model into a topographic, irregular image. The variation of the model is the result of a compromise between anthropological and territorial structure. The various tonal qualities of the landscape are here described by areas with different shades of green. The image is then processed with a Rhinoceros Grasshopper Image Sampler, an

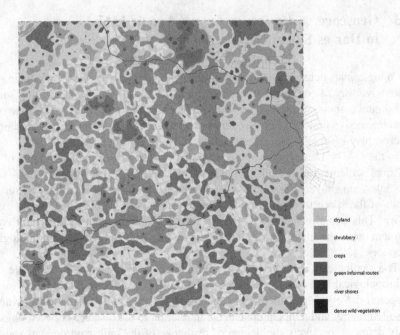

dryland

shrubbery

crops

green informal routes

river shores

dense wild vegetation

Fig. 4.1 Agogic map: ecological layer

application that detects the presence of a higher chromatic concentration where the vegetation is more florid. This procedure constitutes an interpretation and transformation of the available photographic material—the use of satellite data should increase the reliability of the operation, however during the course of this research no satellite images were accessed—the concentrations detected during the image processing are translated into a continuous surface with the highest peaks where the chromatic concentration is higher. The surface is then analyzed and intersected by a series of transversal planes, all at the same distance, in order to define a set of contour curves.

The contour curves are then classified in five qualifications according to five landscape tones defined in regards to the presence of vegetation in the quadrant. We could say that the five levels represent five different conditions of lushness of the vegetation in relation to the presence of built objects. Also, the series establishes a range of possible landscape tones concerning the presence of underground water. The green-grey series is here introduced as a normalization of an environmental gradient that interpolates a natural condition, such as the water level of the soil, with an anthropic settling condition for the exploitation of the same water.

We name this series "green-grey", as it clarifies the topographic relationships between humid ground and the space occupied by buildings or infrastructures. These last two elements represent often a restriction for the natural or simply rural development of these territories. The green-grey map employs five different colors to define such a premise: on the map, the green color represents the areas with the highest concentration of vegetation, the yellow color the areas occupied by build-ings, the green color identifies areas highly populated though with a good presence of vegetation, while the blue color outlines the threshold between green and grey areas—the eco-tone—where the two entities coexist in balance.

The water level on the soil in the quadrant is still quite high, even if the same area is occupied by many dwellings. This is a specific settling mode which invites consideration of the possibility to include bits of wild nature within an integrated urban landscape. The proximity to water basins and infrastructural systems raises an interaction between natural and built environment, identifying with the topography the geographic premise for the construction of the habitat. The water basins and the rivers constitute a set of attractors for the infrastructural system: in fact, they can provide water for the sanitary development of informal settlements, especially during the rainy seasons.

Alluvial basins in Dar es Salaam establish a rhythm for the urban development of the space available on the territory of this metropolis. They define the metrics of transitional areas passing from wet to dry lands. This condition can be fully understood only through the study of the relationship between grey infrastructures and topographic emergencies. Ecotonal areas provide an interesting insight for the morphological analysis of irregular districts. The green-grey series outlines two main attributes of the built environment: the first one, as noted in the previous paragraph, is the correspondence between wet grounds and unplanned urban fabric, and the second one is the complete autonomy of formal infrastructures with respect to topographic emergencies. Both conditions testify to the interdependency existing between the morphological development of informal settlements and the hydric qualification of the ground. Moreover, the presence of water in the soil constitutes a fundamental resource for the survival of unplanned settlements as often they do not receive water supply from the metropolitan system. In some cases, they illegally connect to these plants—weakening their effectiveness—or they exploit unsanitized boreholes [5].

4.5 Accessibility Map

The map of the degree of separation between spatial objects (Fig. 4.2) refers to the building density and the relative proximities between them, drafting the form of the spatial habits of the community. To the distances between different objects corre-spond the level of publicity of the sites and vice versa. There exists an intermediate areal variance where movement is filtered and controlled by local landowners. This map outlines the temporal order of reverberation of connecting elements and infrastructures. In this context, the relationship with the topography qualifies the architectural disposition of the buildings on the ground, establishing a spatial metrics directly deriving from the geographic character of the place. The urban metrics in particular is defined by the distribution of open spaces in the fabric. The image emerging from the system of open spaces is complementary to the map of the

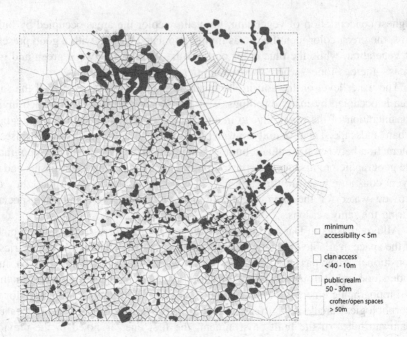

minimum
accessibility < 5m

clan access
< 40 - 10m

public realm
50 - 30m

crofter/open spaces
> 50m

Fig. 4.2 Agogic map: the anthropological layer

tones of the landscape. Even in this case the condition of openness is translated into a type of spatial density described by a surface. This surface is divided by a Voronoi equation, parceling the areas according to a relationship of proximity to the peaks with the highest concentration. The more the peaks are elevated, the greater is the infrastructural hierarchy of the spatial object. The greater is the area contained within each Voronoi, the greater is the level of publicity of the site. We display here a qualitative study of the geometric configuration of the settlement, highlighting emerging relationship between formal and informal contexts.

4.6 A Chronographic Rendition of the Variations of the Model

The environmental map and the map of the degree of separation between spatial objects constitute the two main variables defining what we call the "agogic" of the settlement. They both provide special insight for understanding the morphological variations of the settlement model: proximity to water and vegetation, proximity to buildings. The first two elements refers to the degree of the humidity of the ground and to the fact that the high level of humidity in the soil decreases the resiliency factor of the land, and therefore its livability. The third element indicates the accessibility of a place on the basis of social rules embodied in the configuration of the urban fabric. The agogic map is a map made of three layers: the deepest level is the geometric representation of the settling model, while the other two layers are just layers of variation, the actual agogic of the model (Fig. 4.3).

If the model remains fixed and immutable, or at least if it modifies its config-uration over a long-term basis, the other layers vary according to different speeds, e.g., according to the season cycle, or to accidental demolitions due to flooding or formal urban transformations, or even according to the spatial behavior of the inhabitants. From this point of view, agogic maps are a chronographic rendition of the time of development of the geometric variables of the place on the basis of environmental and social motions.

Working on informal settlements requires an active mapping ability in order to adapt to the fast-changing speeds of the morphological transformation of these spaces. At the same time, this practice requires fixing the character of the place—

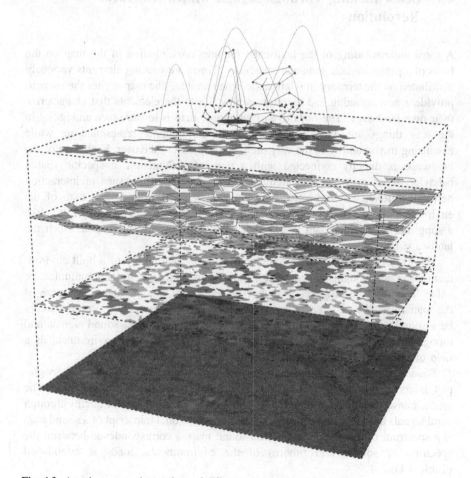

Fig. 4.3 Agogic map: an interactions of different environmental layers

somehow the center of gravity of the mapping system—around which can be drafted a project of internal consolidation in exchange with external, shifting elements. As we noticed, this condition is typical for the city of Dar es Salaam due to the frequent seasonal rains that cause important landscape mutations in the urban region, especially in those degraded, less-resilient areas where the water flow dismantles the weaker fringes of irregular dwellings along alluvial basins. In this context, it is difficult to outline informal settlements as consolidated entities using traditional surveying tools. A cyclical writing of the transformation of the morphology of the settlement should be preferred. The chronographic tracing of the agogic map intertwines space and time in an harmonic field of interaction, according to various modes of absorption and refraction [6].

4.7 Sense-making Through Agogic Maps: An Aural Revolution

A tonal understanding of the landscape requires configuration of the map on the basis of a programmatic semantic instability. Tones are moving elements variously distributed on the territory in relation the time variable. The map invites the users to provide a new meaning and therefore a new life to the elements that characterize their own landscape. The objective of an agogic map is to introduce ambiguity in the way things are written and defined in cartographic representation, while enhancing memorability in the way they are proposed to the user. Each tone of the landscape is directly connected with a tone of inhabitation, a specific spatial behavior. It refers to specific architectural solutions of the issues of interaction between different environments. Each tone traces a threshold. The map of the environmental tones expresses the vocation to resilience of the settlement: It is like a song that every inhabitant knows that recalls the settling rules upon which the landscape is built.

The use of the musical term tone to measure the variability of a built environment leads our thoughts to the aural extent of this entity and to its communicative value. This research proposes a specific declension of "agogic maps" or "maps of the spatial variations". An agogic map based on environmental tones could actually be a sound map if we reconstruct a synesthetic chain between sound signals and topographic characteristics. A sound map exploits an acoustic environment as a field of action for urban reconfiguration and transformation.

Sound cartography not only introduces linguistic mutations in the way geography is communicated, but it enables also innovation of its writing modes. The space, transcribed through the map, is rendered in its experimental quality through aural signals and site-specific metrics. The bi-dimensional transcript of a sound map is a spectrum of these kinetics. In this sound map, a correspondence between the spectrum of sounds and timbres of the environmental tones is established (Table 4.1).

Table 4.1 Agogic map: matrix for the interpolation of sound and space variables

Space variables	Sound variables	
Tone	**Timbre**	**Spectrum**
Dynamic	Intensity	Amplitude
Rhythm	Height	Frequency

As we know, timbre, intensity, and height, are the three variables with which we measure the sound. They correspond to the three variables of motion of the sound wave in space: tone, dynamic, and rhythm. Tone and timbre in particular are included and compared within the spectrum of the sound map. Tones of inhabitation and acoustic timbre are used here to increase the expressivity of each topographic emergency, invite the user to name the sounds encountered.

Eurhythmics in this research is proposed as a synthetic indicator of spatial configuration, according to the anthropological image of the settlement. The tools exploited by Eurhythmics are tone, dynamics, and rhythm [7]. Hugo Riemann was induced to introduce the term "agogic" to indicate all the time mutations that could be manifested, even in the smallest rhythmical combination of sounds. In Riemann's aesthetic vision, the musical pulse is the combination of agogic and tones: "The movement of the human voice in tonic, dynamic and agogic changes, [...] is to be referred to the innate impulse to impart" [7]. In this definition, agogic indicates the variation of a regular model. A work of art, to fully express its artistic potential through the body of who perceives it, has to consider the rhythmical motion of the body that moves around it to be understood. In the agogic definition of space, tones and dynamics are variables that transform the rhythmical monotonous extent of geometric forms into something organic. As happens in music, where the voice provides a large variety of tones and dynamics in order to express a musical work, in space, tones and dynamics collaborate to manifest the inner agogic embodied in the shape of an architecture. In particular, dynamics work on the basis of rhythm and tone. The agogic variation of space elects tone and rhythm as its preferred operators: both of them are implied in the notion of dynamic.

The writing of the space requires the understanding of its expressivity in order to report on the map a realistic rendition of it. Discerning these three operators in relation to the elements that structure the settlement clarifies a qualitative approach to site surveying, describing the place as a combination of various densities. The analyses of the anthropological image of the settlement—a range of numeric values representing tonality, rhythm, and dynamics—is bestowed on each nodal point of the model. This operation enables expression of the agogic of the place, establishing the quality of each site (Fig. 4.4). In particular, the tonality of the place is considered according to the location of the site in relation to its distance to the river or to any other water feature. The rhythm emerges from the extension of the buildings along major and secondary routes. Finally, the dynamic results from the progressive increase or decrease of rhythmical measures approaching each relevant point on the map. The dynamic introduces the speed variable in the process of agogic determination of the field of action of the map. Each point of the map constitutes a cross

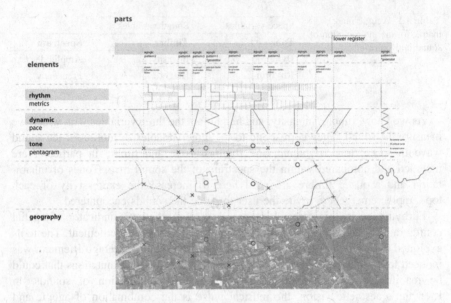

Fig. 4.4 Yombo Vituka: correspondence on the agogic map among values of tone, dynamic, and rhythm

section of the agogic extent of the built environment, describing the kinetic articulation of that specific spatial configuration.

4.8 Spectralism and Spatial Variance

The spectrum of an agogic map is a correlative system (Fig. 4.5). It is a device that builds a bond of interdependency between sound and space variables. This tool intends to translate on a bi-dimensional or tridimensional environment the performative and experimental extents of space. The writing of the expressions creates a relationship between the spectrum of sound signals produced both for the left and the right ear. This approach is applied to a mockup of the device based on a two-minute path around a small portion of the informal settlement of Yombo Vituka. The path intercepts signals traced on the map in the shape of circular areas. Each circle is provided with a value of sound intensity, volume, and tonal height. The sounds applied to the relevant points of the anthropological image of the settlement are consonances (unison-fifth-octave), sounds that are easier to recognize, producing no sense of bewilderment nor disturbance in the users. Along the path, a number of more irritating sound are encountered. These sounds are named dissonances, and they invite the user to acknowledge the state of emergency of those sites and to consider their unstable condition for potential reform and transformation.

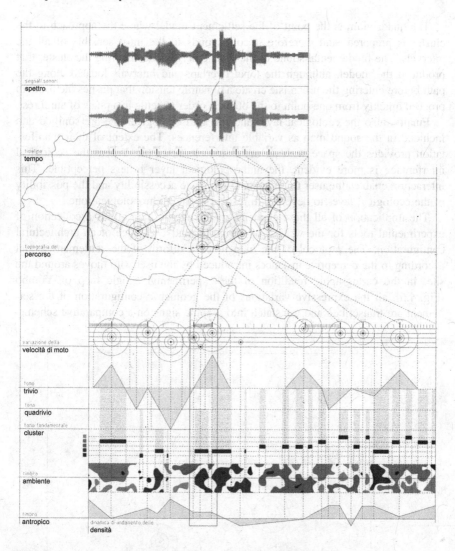

Fig. 4.5 Agogic map: spectrum of dissonance and kinetic variations

Urban pressure and building density constitute an important parameter for the variation of the speed of navigation, therefore affecting the site accessibility. On the stave of Yombo Vituka, consonances and dissonances are placed according to the settlement model; the score is divided into bars following a grid of consistent bars. The speed of navigation resizes the width of each bar according to the user's movement. The various tonal heights of the consonances are connected to the key sites of the anthropological image of the settlement according to the following scheme: trivium—fifth, cluster—unison/octave, quadrivium—seventh.

The quadrivium is the point of the settlement model where the approach to the cluster is prepared, and therefore it corresponds to the most sensible of all the intervals. The fundamental sound of the scale emerges encountering the cluster that produced the model, although the tonal overlaps and intervals located along the path before entering the urban isle create a dynamic tension that pushes the users to proceed quickly from one point to the other in order to settle in a place of aural rest.

Finally, also the ecological layer and the anthropological one are sonified and included in the sound map as variable interferences. The extent of their manifestation provides the space with an undulated character [8]. Where the ecological interference is more evident, the anthropological layer is less perceptible. This interaction enables the user to understand the place accessibility and the possibility of the ecological layer to develop, in respect to the anthropological one.

The implication of all these elements in the agogic map builds a collection of experimental tools for the writing of the performative essence of an architectural configuration. The protocol of the map tunes a reform of the settlement model according to the perceptive instances introduced by the user who moves around the site. In the cartographic rendition of the experimental agogic map of Yombo (Fig. 4.6), all the expressive variations of the geometric configuration of the settlement are transcribed and translated into graphic signs on a comparative scheme.

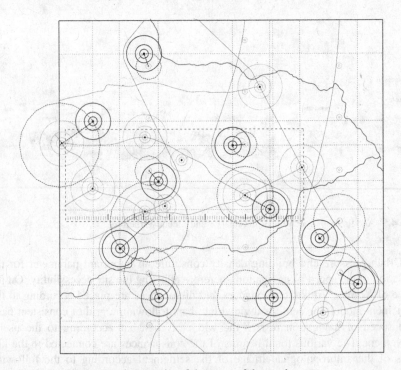

Fig. 4.6 Yombo Vituka: agogic transcript of the space of the quadrant

This operation interprets the agogic of the place as a regulating tool of the curvature of the linear elements of the spatial grid of the settlement model from which emerges the variations of the spatial configuration in relation to topographic, environmental, and anthropological emergencies. This condition implies the tracing of drifts and pace changes while navigating through a built environment.

4.9 Provisional Conclusions

The comparative scheme proposes a comparison between the geographic structure of the settlement model with sound signals. The position of the central cluster that engendered the development of the settlement can be considered in its starting location and confronted with performative deformation. The map offers a rendition of topography as an embodied writing of the expressivity of the space for which an experimental correlation between geography and built environment is fostered and enhanced through a common spatial habit (body, behavior, rhythm). Reading through the words of Paul Valéry, the map is a cuttlefish bone [9]. It is the form in which the life and death of things collide, evoking the antecedent of its essence and, at the same time, outlining the way it will be remembered.

Topography as a tool for spatial orientation expresses the geographic reasons of the structure of a built environment. An agogic map, as a writing system of a sensorial movement in space, combines sound and visual images inviting users to create their own innovative sensorial chains in order to make the place livable and memorable. Agogic maps open our discourse on perceptive topography as a possible discipline to study and to improve the composition and the construction of our habitat. More sensible spatial forms will be conceived as a result of an enhanced care for our environment, for its ecology, and for the infinite expressions of our human behavior.

References

1. Riemann, H. (1884). *Musikalische Dynamik und Agogik*. Leipzig: Breitkopf & Härtel.
2. Chatwin, B. (1987). *The songlines*. London: Franklin Library.
3. Jaques-Dalcroze, E. J. (1920). *Il Ritmo, la Musica, l'Educazione* (p. 148). Torino: EDT.
4. Cfr. McHarg, I. (1967). *Design with nature*. London: Wiley.
5. Cfr. World Bank. (1999). *Tanzania-Dar es Salaam water supply and sanitation project*. Washington DC, USA: World Bank.
6. Cfr. Donatoni, F. (1996). Codici: l'errore come elemento mutageno. In A. Morelli & S. Scarani (Eds.) *Sound design* (pp. 145–146). Bologna: Pitagora. Ciò che nell'estetica donatoniana viene nominato come errore rispetto a una serie certa di suoni, nella mappa agogica l'errore è l'incerto, il non nominato "a priori", quindi il "sensibile" e aperto alla risemantizzazione.

7. Riemann, H. (1895). *Catechism of musical aesthetics* (p. 23). English edition: (trans: Bewerunge, H.), 2013.
8. Cfr. Deleuze, G., Guattari, F. (1980). *Mille Plateaux* (p. 698). Paris: Éditions de Minuit.
9. Cfr. Valéry, P. (1923). *Eupalinos, o l'architetto*. Milano: Bilioteca dell'Immagine (1997).

Bibliography

1. Alberti, L. B. (1991). *On the art of building in ten books*. Cambridge, MA: The MIT Press.
2. Adorno, T. W. (2006). *Philosophy of new music*. Minnesota: Minnesota University Press.
3. Alexander, C. (1964). *Notes on the synthesis of form*. Cambridge, MA: Harvard University Press.
4. Ardalan, N. (1999). *The sense of Unity (Persian Architecture)*. New York: ABC.
5. Arnheim, R. (1969). *Visual thinking*. Los Angeles: University of California Press.
6. Arnheim, R. (1971). *Entropy and art*. Los Angeles: University of California Press.
7. Arnheim, R. (1977). *The dynamics of architectural form*. Los Angeles: University of California Press.
8. Appia, A. (1895). *La mise en scène du drame wagnérien. 1892–1894*. Paris: Léon Challey.
9. Appia, A. (1899). *Die Musik und die Inscenierung*. München: Bruckmann.
10. Appia, A. (1921). *L'Oeuvre d'Art Vivant. 1916–1920*. Genève-Paris: Edition Atar.
11. Bachelard, G. (1957). *La Poétique de l'espace*. Paris: Presses Universitaires de France.
12. Banks, P. (2010). *Represent: Art and identity among the black upper-middle class*. London: Routledge.
13. Bandur, M. (2001). *Aesthetics of total serialism. Contemporary research from music to architecture*. Berlin: Birkhäuser.
14. Barbara, A. (2012). *Sensi, tempo e architettura*. Milano: Postmedia Books.
15. Bateson, G. (1972). *Steps to an ecology of mind*. Chicago: University of Chicago Press.
16. Bateson, G. (1991). *Sacred unity: Further steps to and ecology of mind*. Lyme CT: Cornelia & Bessie Foundation.
17. Batty, M. (2005). *Cities and complexity. Understanding the cities with cellular automata, agent-based models, and fractals*. Cambridge, MA: MIT Press.
18. Baudrillard, J. (1981). *Simulacres et simulation*. Paris: Galilée.
19. Baudrillard, J. (2003). *Mass, identity, architecture: Architectural writings*. Chichester: Wiley.
20. Bergson, H. (2001). *Time and free will: An essay on the immediate data of consciousness*. London: Dover Books.
21. Bertin, J. (1967). *Semiologie grafique*. Paris: Mouton.
22. Berio, L. (1956). Prospettive nella musica. Ricerche ed attività dello Studio di Fonologia Musicale di Radio Milano. *Elettronica, 3*, 108–115.
23. Berio, L. (1981). *Intervista sulla musica*. Milano: Laterza.
24. Berio, L. (2006). *Un ricordo al futuro (Lezioni Americane)*. Torino: Einaudi.
25. Boethius, S. (1990). *The consolations of music, logic, theology, and philosophy (Clarendon Paperbacks)*. Oxford: Oxford University Press.
26. Bossini, O. (2009). *Milano, laboratorio musicale del Novecento. Scritti per Luciana Pestalozza*. Milan: Archinto.
27. Boulez, P. (1963). *Penser la musique aujourd'hui*. Paris: Tel Gallimard.

© The Author(s) 2017
R. Pe, *Agogic Maps*, PoliMI SpringerBriefs,
DOI 10.1007/978-3-319-48306-1

28. Brambati, R. (2011). Écologie sonore et technologie du son. In *Sonorités. Écologie Sonore Technologies Musiques* (pp. 9–41), n° 6 September 2011. Paris: Champ Social.
29. Branzi, A. (2006). *Weak and diffuse modernity*. Milano: Skira.
30. Bunschoten, R. (1999). *Stirring the City*. Berlin: Fonden til Udgivelse af Arkitekturtidsskrift.
31. Bunschoten, R. (2002). *Public Spaces*. London: Black Dog.
32. Bunschoten, R. (2005). *CHORA: From matter to metaspace*. Wien: Springer.
33. Bunschoten, R. (2006). Touching the second skin. In K. Oostheruis & L. Feireiss (Eds.), *Game, set and match: No.2: The architecture of co-laboratory* (pp. 598–611). Rotterdam: Episode.
34. Cage, J. (1969). *Notations*. New York: Something Else Press.
35. Camillus, J. S. (1994). *Urban Agriculture and the Sustainable Dar-es-Salaam Project*. CFP Report 10—University of Dar es Salaam, Dar es Salaam, United Republic of Tanzania.
36. Carpo, M. (2011). *The Alphabet and the algorithm*. New York: MIT Press.
37. Carratelli, C. (2006). *L'integrazione dell'estesico L'intégration de nel poietico nella poetica musicale post-strutturalista. Il caso di Salvatore Sciarrino. Una composizione dell'ascolto.* Università degli Studi di Trento, Tesi di Dottorato, Trento.
38. Cattaneo, C. (1858). *La città considerate come principio ideale delle storie italiane*. Torino: Einaudi.
39. Chatwin, B. (1987). *The songlines*. London: Franklin Library.
40. Choay, F. (1969). *The modern city: Planning in the Nineteenth Century*. New York: Braziller.
41. Contin, A., Paolini, P., & Salerno, R. (2014). *Innovative technologies in urban mapping, built space and mental space*. Milano: Springer.
42. Contin, A. (2015). *Questo. Metropolitan Architecture*. Santarcangelo di Romagna: Maggioli.
43. Cosgrove, D. (1989). *Social formation and symbolic landscape*. London: Crum Helm.
44. Cosgrove, D. (1999). *Mappings*. London: Reaktion Books.
45. Crespi Morbio, V. (2013). *Appia alla Scala*. Milano: Ricordi.
46. D'Alfonso, E. (1997). *Intrecci spaziotemporali dell'architettura*. Milano: Electa.
47. Dalcroze, E. J. (1906). *Les gammes, les tonalités, le phrasé & les nuances*. Paris: Sandoz-Jobin.
48. Dalcroze, E. J. (1916). *La rythmique, la plastique animée et la danse*. Lausanne: Jobin & Cie.
49. Dalcroze, E. J. (2011). *Le Rythme, La Musique Et L'éducation*. Paris: Nabu Press.
50. Dalcroze, E. J. (1945). *La musique et nous*. Genève-Paris: Slatkine.
51. Dar es Salaam City Commission. (1998). *Strategic urban development plan for Dar es Salaam, draft plan on city growth/expansion, vision and pattern*. Dar es Salaam, Tanzania: Sustainable Cities Programme.
52. DAWASA (Dar es Salaam Water and Sewerage Authority). (2000). *Management Report (technical report covering the period October–December 1999), Internal Management Report*. Dar es Salaam, United Republic of Tanzania: DAWASA.
53. De Martino, E. (1958). *Morte e pianto rituale nel mondo antico: dal lamento pagano al pianto di Maria*. Torino: Bollati Boringhieri.
54. Dematteis, G. (1970). *«Rivoluzione qunatitativa» e nuova geografia*. Torino: Università di Torino.
55. Dematteis, G. (1985). *Le Metafore della Terra: la geografia umana tra mito e scienza*. Milano: Feltrinelli.
56. Derrida, J. (1972). *Positions*. Paris: Minuit.
57. Deleuze, G. (1988). *Le pli. Leibniz et le Baroque*. Paris: Éditions de Minuit.
58. Deleuze, G., & Guattari, F. (1972). *L'Anti-Oedipe*. Paris: Éditions de Minuit.
59. Deleuze, G., & Guattari, F. (1980). *Mille Plateaux*. Paris: Éditions de Minuit.
60. Deleuze, G., Guattari, F. (1983). *Rhizome*. New York: Semiotext(e).

61. Dewey, J. (2004). *Essays in experimental logic*. London: Dover Publications.
62. Debord, G. (1967). *La Société du spectacle*. Paris: Èditions Buchet-Chastel.
63. Diaz, Olvera L., Plat, D., & Pochet, P. (2003). Transportation conditions and access to services in a context of urban sprawl and deregulation. The case of Dar es Salaam. *Transport Policy, 10,* 287–298.
64. Donatoni, F. (1970). *Questo*. Milano: Adelphi.
65. Donatoni, F. (1980). *Antecedente X*. Milano: Adelphi.
66. Donatoni, F. (1996). Codici: l'errore come elemento mutageno. In A. Morelli & S. Scarani (Eds.), *Sound design* (pp. 145–146). Bologna: Pitagora.
67. Dorfles, G. (1996). *Il divenire delle arti*. Milano: Bompiani.
68. Dorita, H., & Harslof, O. (2008). *Performance design*. Copenhagen: Narayana Press.
69. Duchamp, M. (1959). *Marchand de sel: Ecrits de Marcel Duchamp*. Paris: Sanouillet.
70. Eco, U. (1962). *Opera aperta*. Milano: Bompiani.
71. Eco, U. (2012). *Scritti sul Pensiero Medievale*. Milano: Bompiani.
72. Eisenman, P. (1998). *Blurred zones: Investigations of the interstitial: Eisenman Architects 1988–1998*. New York: Monacelli Press.
73. Eisenman, P. (2003). *Giuseppe Terragni: Transformations, decompositions, critiques by Peter Eisenman*. New York: The Monacelli Press.
74. Eisenman, P. (2008). *Ten canonical buildings 1950–2000*. New York: Rizzoli.
75. Eisenman, P. (2009). *The formal basis of modern architecture*. New York: Lars Muller.
76. Farldi, G. (2010). *Valutazione della vulnerabilità al cambiamento climatico delle comunità costiere di Dar es Salaam (Tanzania) rispetto al fenomeno dell'intrusione salina nella falda acquifera*. Roma: Università della Sapienza.
77. Farina, A., & Belgrano, A. (2006). The eco-field hypothesis: Toward a cognitive landscape. *Landscape Ecology, 21,* 5–17.
78. Farinelli, F. (2003). *Geografia*. Torino: Einaudi.
79. Favaro, R. (2010). *Spazio Sonoro. Musica e architettura tra analogie, riflessi, complicità*. Venice: Marsilio.
80. Focillon, H. (1989). *The life of forms in art*. London: Zone Books.
81. Formaggio, D. (1990). *Estetica Tempo Progetto*. Milano: Città Studi.
82. Franzini, E. (1991). *Metafora, Mimesi, Morfogenesi, Progetto*. Milano: Guerini.
83. Franzini, E. (1994). *Arte e mondi possibili. Estetica e interpretazione da Leibniz a Klee*. Milano: Guerini.
84. Franzini, E. (2001). *Fenomenologia dell'invisibile. Al di là dell'immagine*. Milano: Cortina.
85. Franzini, E. (2008). *I simboli e l'invisibile. Figure e forme del pensiero simbolico*. Milano: Il Saggiatore.
86. Gadamer, H.-G. (1986). *The relevance of the beautiful*. New York: Bernasconi.
87. Gaube, H. (2007). *Der frühe Islam*. Berlin: Schiler Verlag.
88. Gombrich, E. (1986). *Aby Warburg. An intellectual biography*. Chicago: University of Chicago.
89. Gregotti, V. (2010). *Architecture, Means and Ends*. Chicago: University of Chicago.
90. Greimas, A. J. (1984). *Structural semantics: An attempt at a method*. University of Nebraska Press.
91. Grunebaum, G. (1962). *Modern Islam: The search for cultural identity*. Los Angeles: Berkley University Press.
92. Guattari, F. (1995). *Chaosmosis: An ethico-aesthetic paradigm* (trans: Bains, P., & Pefanis, J.). Sydney: Power Publications.
93. Hall, E. (1990). *The hidden dimension*. London: Anchor.
94. Harper, A. (2011). *Infinite music*. Alresford: John Hunt Publishing.
95. Harris Stahl, W. (1993). *Martianus Capella and the Seven Liberal Arts (Records of Western Civilization Series)*. New York: Columbia University Press.
96. Heidegger, M. (1977). *On time and being*. New York: Cotler Books.
97. Heidegger, M. (1964). *Corpo e Spazio*. Genova: Melangolo.

98. Henderson, L. (1983). *The fourth dimension and non-Euclidean geometry in modern art.* Princeton: Princeton University Press.
99. Henderson, L. (2002). *From energy to information: Representation in science and technology, art, and literature.* Stanford: Stanford University Press.
100. Hillier, B. (1993). *The social logic of space.* London: Bell & Bain.
101. Hillier, B. (1998). *Space is the machine.* Cambridge: Cambridge University Press.
102. Hourani, A. (2013). *A history of the Arab Peoples.* Cambridge: Cambridge University Press.
103. Husserl, E. (1982). *Ideas pertaining to a pure phenomenology and to a phenomenological philosophy: First book: General introduction to a pure phenomenology.* Berlin: Springer.
104. Husserl, E. (1997). *Thing and space: Lectures of 1907.* Berlin: Springer.
105. Kironde, J. M. (1994). *The evolution of land use structure of Dar es Salaam 1890–1990. A study in the effects of land policy.* Ph.D. Thesis, University of Nairobi, Nairobi, Kenya.
106. Kironde, J. M. L. (1997). Towards efficient urban land use in Tanzania. In W. Kombe & V. Kreibich (Eds.), *Urban land management and transition to market economy in Tanzania* (pp. 59–68). SPRING Research Series No. 19, Dortmund.
107. Kjellen, M. (2006). *From public pipes to private hands. Water access and distribution in Dar es Salaam, Tanzania.* Ph.D. Thesis, Department of Human Geography, Stockholm University, Intellecta DocuSys AB, Solna, Sweden.
108. Kyessi, A. G. (2002). *Community participation in urban infrastructure provision. Servicing informal settlements in Dar es Salaam, Tanzania.* Ph.D. Thesis, University of Dortmund, Dortmund, Germany.
109. Klee, P. (1925). *Pädagogisches Skizzenbuch.* Weimar: Neue Bauhausbücher; Kombe, W. J. (1995). *Formal and informal land management in Tanzania, the case of Dares Salaam.* SPRING Research Series No. 13, Dortmund.
110. Kombe, W. J. (1999). Urban poverty, service provision and land development in Tanzania: Predicaments and prospect. *Report of Workshop Proceedings* (pp. 6–27). Arusha, Tanzania: Institute of Engineers of Tanzania (IET).
111. Kombe, W. J., & Kreibich, V. (2000). *Formal land management in Tanzania.* SPRING Research Series No. 29, Dortmund.
112. Kombe, W. J. (2005). Land use dynamics in peri-urban areas and their implication on urban growth and form: The case of Dar es Salaam, Tanzania. *Habitat International, 29*(1), 113–135.
113. Jacod, J. (2000). *Probability essentials.* Berlin: Springer.
114. Jankélévitch, V. (1961). *La Musique et l'Ineffable.* Paris: Colin.
115. Latour, B. (1987). *Science in action.* Cambridge: Harvard University Press.
116. Latour, B. (1991). Technology is society made durable. In *A sociology of monsters.* London: Routledge.
117. Le Corbusier, (1991). *Precisions: On the present state of architecture and city planning.* Cambridge, MA: The MIT Press.
118. Le Corbusier, (1923). *Vers une Architecture.* Paris: Cres.
119. Le Corbusier, (1948). *Le Modulor.* Boulogne-Billancourt: Éditions de l'Architecture d'aujourd'hui.
120. Leatherbarrow, D. (2004). *Topographical stories: Studies in landscape and architecture.* Philadelphia US: Pennsylvania University Press.
121. Lefebvre, H. (1974). *La production de l'espace.* Paris: Anthropos.
122. Lembi, P., & Moro, A. (2010). *Esperienze dello Spazio.* Milano: Politecnica.
123. Lévi-Strauss, C. (1955). *Tristes tropiques.* London: Penguin Modern Classics.
124. Lévi-Strauss, C. (1958). *Anthropologie Structurale.* Paris: Pocket.
125. Lévi-Strauss, C. (1962). *Le totémisme aujourd'hui.* Paris: Presses Universtiaires de France.
126. Lévi-Strauss, C. (1962). *La Pensée sauvage.* Paris: Presses Universtiaires de France.
127. Lévi-Strauss, C. (1978). *Mythologiques - Le cru et le cuit.* Paris: Plon.

128. Lupala, J. M. (2002). *Urban types in rapidly urbanizing cities: A typological approach in the analysis of urban types in Dar es Salaam*. Ph.D. Thesis, Department of Infrastructure, Division of Urban Studies, Royal Institute of Technology, Stockholm, Sweden.

129. Lupala, A. (2002). *Peri-urban land management for rapid urbanization. The case of Dar es Salaam*. SPRING Research Series 32, Dortmund, Germany.

130. Lynch, K. (1964). *The image of the city*. Cambridge, US: MIT Press.

131. Lynch, K. (1964). *The view from the road*. Cambridge, US: MIT Press.

132. Marsciani, F. (2007). *Traccati di etnosemiotica*. Milano: Franco Angeli.

133. Mato, R. R. A. M. (2002). *Groundwater pollution in urban Dar es Salaam, Tanzania: Assessing vulnerability and protection priorities*. Ph.D. Thesis, Eindhoven University of Technology, University Press, Eindhoven, Netherlands.

134. Mbonile, M. J., & Kivelia, J. (2008). Population, environment and development in Kinondoni Districts, Dar es Salaam. *Geographical Journal, 174*(2), 149–175.

135. McGee, T. G. (1967). *The Southeast Asian city, London*. London: Bell and Sons, Ltd.

136. McGee, T. G. (1971). *The urbanization process in the third world*. London: Bell and Sons.

137. McHarg, I. (1967). *Design with nature*. London: Wiley.

138. McHarg, I. (1970). *Conversations with students, Dwelling in Nature*. New York: Princeton Press.

139. McLuhan, M., & Fiore, Q. (1967). *The Medium is the massage*. New York: Bantam.

140. McLuhan, M. (1969). *Understanding media*. London: Routledge.

141. Merleau-Ponty, M. (1945). *Phénoménologie de la perception*. Paris: Éditions Gallimard.

142. Merleau-Ponty, M. (1966). *Sens et non-sens*. Paris: Éditions Nagel.

143. Merleau-Ponty, M. (1979). *Le visible et l'invisible*. Paris: Éditions Gallimard.

144. Mersenne, M. (2013). *Harmonie universelle, contenant la théorie et la pratique de la musique* (Vol. I, II). Paris: Hachette Livre BNF.

145. Mjemah, I. C. (2007). *Hydrogeological and hydrogeochemical investigation of a coastal aquifer in Dar es Salaam, Tanzania*. Ph.D. Thesis, Ghent University, Ghent, Belgium.

146. Mila, M. (1976). *Maderna musicista europeo*. Torino: Einaudi.

147. Mosi, J. B. R. (1996). *Urban population growth and accessibility to domestic water supply in Tanzania. A case study of Dar es Salaam*. Dar es Salaam: University of Dar es Salaam.

148. Msindai, K. (1988). *Engineering geological aspects of soil and rocks in the Dar-es-Salaam region, Tanzania*. Ph.D. Thesis, Institute of Quaternary Geology, University of Turku, Turku, Finland.

149. NBS (National Bureau of Statistics). (2002). *Population and housing census*. Dar es Salaam, United Republic of Tanzania.

150. Norberg-Shultz, C. (1988). *Architecture, Meaning and Place*. New York: Rizzoli.

151. Jankélévitch, V. (1961). *La Musique et l'Ineffable*. Paris: Colin.

152. Kanach, S. (2001). *Music and architecture by Iannis Xenakis*. New York: Pendragon Press.

153. Kandinski, V. (1945). *On the spiritual in art*. New York: The Guggenheim Foundation.

154. Koolhaas, R. (1997). *Delirious New York*. New York: Monacelli Press.

155. Koolhaas, R. (2006). *S, M, L, XL*. New York: Monacelli Press.

156. Koolhaas, R. (2006). *Junkspace*. Macerata: Quodlibet.

157. Ortiz, P. (2013). *The art of shaping metropolis*. Boston: McGraw-Hill.

158. Paci, E. (1947). *Esistenza e imagine*. Milano: Tarantola.

159. Paci, E. (1954). *Tempo e relazione*. Torino: Taylor.

160. Paci, E. (1961). *Diario fenomenologico*. Milano: Il Saggiatore.

161. Paci, E. (1966). *Relazioni e significati (tre volumi)*. Milano: Lampugnani Nigri.

162. Panofski, E. (1991). *Perspective as symbolic form*. Cambridge, MA: The MIT Press.

163. Panofski, E. (2005). *Gothic architecture and scholasticism*. London: Archabbey Publications.

164. Peirce, C. S. (1868). Questions concerning certain facoulties claimed for man. In *Collected Papers*. Cambridge: Harvard University Press.

165. Peirce, C. S. (1893). *Of Reasoning in general*. In *Collected Papers*. Cambridge: Harvard University Press.
166. Peirce, C. S. (1895). *What is a sign?*. In *Collected Papers*. Cambridge: Harvard University Press.
167. Peirce, C. S. (1923). *Chance, Love, and Logic*. New York: Brace.
168. Peirce, C. S. (1931–58). *Collected Papers of Charles Sanders Peirce*. Cambridge: Harvard University Press.
169. Peirce, C. S. (1977). *Semiotic and significs*. Indianapolis: Indiana University Press.
170. Planck, M. (1965). *Vortrage Und Erinnerungen*. Berlin: Wissenschaftliche.
171. Poincaré, J. H. (1929). *La Science et l'Hypothèse*. Paris: Flammarion.
172. Portugali, J. (2000). *Self-organization and the city*. Berlin: Springer.
173. Reimann, H. (1884). *Musikalische Dynamik und Agogik*. Leipzig: Breitkopf & Härtel.
174. Reimann, H. (1908). *Handbuch der Harmonielehre*. Leipzig: Breitkopf & Härtel.
175. Reimann, H. (1919). *Musik-Lexikon*. Leipzig: Breitkopf & Härtel.
176. Reimann, H. (1920). *Lehrbuch des Contrapunkts*. Leipzig: Breitkopf & Härtel.
177. Ricoeur, P. (1965). Universal civilization and national cultures. In *History and truth*. Evanston: Northwestern University Press.
178. Rykwert, J. (2008). *The judicious eye. Architecture against the other arts*. London: Reaktion Book.
179. Ricci, L. (2010). *Assessing the local adaptive capacity to climate change: A pilot study to understand the autonomous environmental management in Dar es Salaam*. 20th Conference INURA (International Network for Urban Research and Action), "The New Metropolitan Mainstream", Zürich, Switzerland, June 27–July 3, 2010.
180. Rogers, E. N. (1958). *Esperienza dell'architettura*. Torino: Einaudi.
181. Rogers, E. N. (1980). *Gli elementi del fenomeno architettonico*. Milano: Mariotti.
182. Rognoni, L. (1974). *Fenomenologia della musica radicale*. Milan: Aldo Garzanti Edizioni.
183. Rossi, A. (1966). *L'Architettura della Città*. Padova: Marsilio.
184. Rowe, C., & Koetter, F. (1978). *Collage city*. New York: MIT Press.
185. Rowe, C. (1982). *The mathematics of the ideal villa and other essays*. Cambridge, MA: The MIT Press.
186. Sartre, J. P. (1940). *L'imaginaire*. Paris: Solio Dessais.
187. Sauvaget, J. (1965). *Introduction to the history of Muslim East*. Los Angeles: Berkeley University Press.
188. Sawio, C. J. (2008). Perception and conceptualization of urban environmental change: Dar es Salaam City. *Geographical Journal, 174*(2), 149–175.
189. Schlemmer, O. (1990). *The letters and diaries of Oskar Schlemmer*. The Northwestern University Press.
190. Schönberg, A. (1983). *Theory of harmony*. Los Angeles: California University Press.
191. Schumacher, P. (2012). *The Autopoiesis of Architecture* (Vol. I, II). London: Wiley.
192. Sciarrino, S. (1985). *L'immagine del suono (grafici 1966–1985)*. Le Batiment Deux: Latina.
193. Sciarrino, S. (2001). *Carte da suono (1981–2001)*. Roma-Palermo: Cidim/Novecento.
194. Shane, D. G. (2011). *Urban design since 1945: A global perspective*. London: Wiley.
195. Sini, C. (1991). *L'espressione e il profondo*. Milano: Lanfranchi.
196. Sini, C. (1992). *Pensare il progetto*. Milano: Tranchida.
197. Sini, C. (1993). *L'incanto del ritmo*. Milano: Tranchida.
198. Sini, C. (1994). *Filosofia e scrittura*. Roma-Bari: Laterza.
199. Sini, C. (1996). *Gli abiti, le pratiche, i saperi*. Milano: Jaca Book.
200. Shepard, M. (2011). *Sentient city*. Cambridge, US: MIT Press.
201. Smithson, R. (1969). *Incidents of mirror-travel in the Yucatan*. California: Artforum VIII.
202. Smithson, R. (1996). *The collected writings*. Berkeley: University of California Press, San Francisco.
203. Stefani, G. (1982). *Il linguaggio della musica*. Roma: Paoline.
204. Stefani, G. (1987). *Il segno della musica. Saggi di semiotica musicale*. Palermo: Sellerio.

205. Sudgen, S. (2007). *Excreta management in unplanned areas*. London: London School of Hygiene and Tropical Medicine.
206. Tafuri, M. (1973). *Progetto e Utopia*. Bari: Laterza.
207. Thom, R. (1994). *Structural stability and morphogenesis*. New York: The Westview Press.
208. Tschumi, B. (1983). *The discourse of events*. London: Architectural Association.
209. Tschumi, B. (1996). *Architecture and disjunction*. Cambridge, US: MIT Press.
210. UNDP (United Nations Development Programme). (2007). *Lo sviluppo umano, rapporto 2007/2008. Resistere al cambiamento climatico* (titolo originale: Human Development Report 2007/2008. Fighting climate change: Human solidarity in a divided world). Torino, Italia: Rosenberg & Sellier.
211. UNESCO (United Nations Educational, Scientific and Cultural Organization) Commissione Italiana. (2007). *Sviluppo sostenibile e cambiamenti climatici*. Roma, Italia: Tipolitografia Trullo.
212. UN-HABITAT (The United Nations Human Settlements Programme), UNEP (United Nations Environment Programme). (2004). *The sustainable Dar es Salaam Project 1992–2003. From urban environment priority to up-scaling strategies city-wide*. Nairobi, Kenya: UNION, Publishing Services Section.
213. UN-HABITAT (The United Nations Human Settlements Programme). (2009). *Tanzania: Dar es Salaam City Profile*. Nairobi, Kenya: UNION, Publishing Services Section.
214. UN-HABITAT (The United Nations Human Settlements Programme) (2). (2009). *Cities and climate change initiative, Launch and Conference Report, Oslo, Norway, 17/03/2009*. Nairobi, Kenya: UNION, Publishing Services Section.
215. UN-HABITAT (The United Nations Human Settlements Programme). (2010). *Citywide action plan for upgrading unplanned and unserviced settlements in Dar es Salaam*. Nairobi, Kenya: UNION Publishing Services Section; WaterAid. (2003). *Water reforms and PSP in Dar es Salaam. New rules: Does PSP benefit the poor?*, WaterAid—Tanzania, WaterAid and Tearfund, Dar es Salaam.
216. Vidler, A. (1994). *The architectural uncanny: Essays in the modern unhomely*. Cambrige, MA: The MIT Press.
217. Varela, F., & Maturana, H. (1980). *Autopoiesis and cognition: The Realization of the Living*. Boston: Reidel.
218. Varela, F., Thompson, E., & Rosch, E. (1991). *The embodied mind: Cognitive science and human experience*. Boston: MIT Press.
219. Virilio, P. (1993). *L'espace critique*. Paris: Bourgois.
220. Warburg, A. (1988). *A lecture on serpent ritual*. London: The Warburg Institute.
221. Wardhaugh, B. (2014). The Compendium Musicæ of René Descartes: Early English Responses (Musical Treatises). Brepols Publisher.
222. WaterAid. (2005). *Water and sanitation in Tanzania: An update based on the 2002 Population and Housing Census*. London: WaterAid.
223. Webern, A. (2014). *The path to new music*. New York: Nabu Press.
224. Wyshogrod, E. (2000). *Emmanuel Levinas: The problem of ethical metaphysics*. New York: Fordham University Press.
225. Wittkower, R. (1971). *Architectural principles in the age of humanism*. London: Norton & Company.
226. World Bank. (1999). *Tanzania-Dar es Salaam Water Supply and Sanitation Project*. Washington DC, USA: World Bank.
227. Xenakis, I. (1976). *Musique architecture*. Paris: Casterman.
228. Xenakis, I. (1963). Musique formelle. In «*Revue musicale*», n. 253-4. Paris: Masse.
229. Xenakis, I. (1971). *Formalized music. Thought and mathematics in composition*. London: Bloomington Press.

230. Xenakis, I. (1976). Arti/Scienze: Leghe. In «*Quaderni della Civica Scuola di Musica*», n. 18, Dicembre 1989.
231. Xenakis, I. (1996, tr. It. 1982). *Musique et originalité*. Paris: Séguier.
232. Yates, F. (1966). *The art of memory*. Chicago: University of Chicago.

Websites

233. http://www.everydaylistening.com/articles/2009/12/15/cymatics-visualizing-sound.html
234. http://ambientalternity.tumblr.com
235. http://www.ted.com/talks/david_byrne_how_architecture_helped_music_evolve.html
236. http://www.ted.com/talks/lang/tr/michael_pawlyn_using_nature_s_genius_in_architecture.html
237. http://www.ted.com/talks/lang/tr/alain_de_botton_atheism_2_0.html
238. http://www.ted.com/talks/honor_harger_a_history_of_the_universe_in_sound.html
239. http://www.laboralcentrodearte.org/en/exposiciones/visualizar-el-sonido
240. http://www.youtube.com/watch?v=05Io6lop3mk
241. http://www.ted.com/talks/evelyn_glennie_shows_how_to_listen.html
242. http://cymatica.net
243. http://www.labiennale.org/it/mediacenter/video/iannix.html